Sustainable Civil Infrastructures

Sustainable Infrastructure impacts our well-being and day-to-day lives. The infrastructures we are building today will shape our lives tomorrow. The complex and diverse nature of the impacts due to weather extremes on transportation and civil infrastructures can be seen in our roadways, bridges, and buildings. Extreme summer temperatures, droughts, flash floods, and rising numbers of freeze-thaw cycles pose challenges for civil infrastructure and can endanger public safety. We constantly hear how civil infrastructures need constant attention, preservation, and upgrading. Such improvements and developments would obviously benefit from our desired book series that provide sustainable engineering materials and designs. The economic impact is huge and much research has been conducted worldwide. The future holds many opportunities, not only for researchers in a given country, but also for the worldwide field engineers who apply and implement these technologies. We believe that no approach can succeed if it does not unite the efforts of various engineering disciplines from all over the world under one umbrella to offer a beacon of modern solutions to the global infrastructure. Experts from the various engineering disciplines around the globe will participate in this series, including: Geotechnical, Geological, Geoscience, Petroleum, Structural, Transportation, Bridge, Infrastructure, Energy, Architectural, Chemical and Materials, and other related Engineering disciplines.

More information about this series at http://www.springer.com/series/15140

Hesham Ameen · Tamer Sorour
Editors

Sustainability Issues in Environmental Geotechnics

Proceedings of the 2nd GeoMEast International Congress and Exhibition on Sustainable Civil Infrastructures, Egypt 2018 – The Official International Congress of the Soil-Structure Interaction Group in Egypt (SSIGE)

Springer

Editors
Hesham Ameen
Housing and Building National Research Center
Cairo, Egypt

Tamer Sorour
Ain-Sham University
Cairo, Egypt

ISSN 2366-3405 ISSN 2366-3413 (electronic)
Sustainable Civil Infrastructures
ISBN 978-3-030-01928-0 ISBN 978-3-030-01929-7 (eBook)
https://doi.org/10.1007/978-3-030-01929-7

Library of Congress Control Number: 2018957099

This Springer imprint is published by the registered company Springer Nature Switzerland AG
The registered company address is: Gewerbestrasse 11, 6330 Cham, Switzerland

Contents

About the Editors

Hesham Ameen
Geo-Institute of HBRC is considered one of the largest geotechnical institutes in the Middle East that has ISO certificate. Professor Hisham had been selected to chair the institute many years ago. This selection had been done based on his activities, accuracy, awareness, and hard work in the research. He is specialized in geotechnical testing, neural networks for the geotechnical engineering, and soil improvement.

Tamer Sorour

Assistant Professor Tamer Sorour, Faculty of Engineering, Ain Shams University, Cairo, Egypt

Assistant Professor Tamer Sorour is a Geotechnical Engineer graduated from the Faculty of Engineering, Ain Shams University, and obtained his PhD from the Faculty of Engineering, Ain Shams University in 2009. He performed various researches in different fields of geotechnical engineering, such as soil dynamics including the hazards accompanied with pile driving in different soil formations as well as the design of machine foundations and soil liquefaction, and the stability of soil slopes under the effect of earthquakes. The field of deep excavations includes different shoring and anchorage systems. The field of soil improvement using stone columns and the effect of the installation technique on the behavior of the stone columns. He also shared in different researches that are focused on the response of large-diameter piles in different soil formations. He also have minor/major contribution in different mega projects in Egypt as supervisor of structural and geotechnical works including but not limited to: geotechnical and geo-environmental studies, execution of boreholes, shallow and deep excavation works, performing different laboratory testing, field control and field testing using different techniques, installation of different types of piles, retaining walls and different supported deep excavation systems, soil stabilization, dewatering, protection of side slopes and landslides, grouting, and field monitoring.

Effect of Density and Carbonates on Thermal Properties of Virgin Soil

Vanshika B. Muchhara[1(✉)], Lalit S. Thakur[2], and Avani R. Umatt[3]

[1] Civil Engineering Department, Parul University, Vadodara, Gujarat, India
vanshika.muchhara@paruluniversity.ac.in
[2] Civil Engineering Department, Babaria Institute of Technology, Vadodara, Gujarat, India
lalitthakur.cv@bitseducampus.ac.in
[3] Department of Health, Life and Applied Sciences, TeamLease Skill University, Vadodara, India
avani.u@teamleaseuniversity.ac.in

Abstract. Thermal properties of the soil gain great importance in engineering projects and situations where heat transfer may take place and affect the soil properties on the whole. Soil gets affected, for example, during laying of bituminous roads, airfield strips, gas or steam pipelines and/or hot water and cold gas lines in unfrozen grounds. In-situ determination of thermal properties of soil is not only time consuming, but in certain cases, costly. The purpose of this study is to understand the variation in thermal properties of virgin soil with addition of different carbonates like potassium carbonate, sodium carbonate and ammonium carbonate using laboratory based Lee's and Charlton's apparatus. It is observed that increase in density of soil with any of the three chemicals increases the thermal conductivity, however with the selected chemical the increase is limited to a certain percentage addition after which it does not show any changes as compared to virgin soil.

1 Introduction

Soil is a natural body comprised of solids (minerals and organic matter), liquid, and gases that occurs on the land surface, occupies space, and is characterized by one or both of the following: horizons, or layers, that are distinguishable from the initial material as a result of additions, losses, transfers, and transformations of energy and matter or the ability to support rooted plants in a natural environment. The most important realization of present time is that soils must be regarded as system with interacting and independent physical, chemical and biological components. This is exemplified in study of various properties of soils, specifically water quality and thermal issues. The fate of substance which has the potential to be transported from near the soil surface region to the water table is influenced by chemical and microbial transformations, movement between liquid phase and solid phase through sorption reactions, volatilization & uptake by plants. These processes are in turn influenced by characteristics of soil such as mineralogical and organic composition and temperature.

H. Ameen and T. Sorour (Eds.): GeoMEast 2018, SUCI, pp. 1–14, 2019.
https://doi.org/10.1007/978-3-030-01929-7_1

With the preceding sentiment as motivation the present study involved the development of a measurement method for soil thermal properties along with a means to develop a mechanism to examine the interaction between the thermal and hydraulic aspects of a given soil system.

Lee's and Charlton's apparatus is based on the principle that the steady state rate of heat flowing into a system being equal to rate of heat flowing out of the system. Here the rate of heat conducted by poor conductor to the Lee's disc below it is equated to rate of heat radiated by the Lee's disc itself.

1.1 Material Used

The present study was carried out on three types of soils i.e. expansive soils, clayey silt and a single mineral soil commercially available soil i.e. Kaolinite. A single mineral soil was used to remove the heterogeneity due to various minerals in naturally occurring soils. The expansive soil (henceforth referred to as Black Cotton Soil) used for the study was collected from Parekha Village, Kayavaran (Coordinates: 22.067°N 73.2500°E). The clayey silt (henceforth referred to as Silt) was collected from the Faculty of Technology & Engineering. MSU premises (Coordinates: 22.2979°N, 73.1967°E). After removing all organic matter, both these soils were oven-dried, pulverized and then sieved through 425 micron sieves prior to use. Physical and engineering properties of the soils were evaluated and tabulated in Table 1.

Table 1. Index properties of soil

Type of soil	LL* (%)	PL* (%)	SL* (%)	FSI* (%)	Classification	MDD (gm/cc)	OMC (%)
Kaolinite	56	31	16	25	CI	1.48	26.91
BC Soil	56	27	11	75	CH	1.64	18.42
Silt	30	21	21.6	11	MI	1.72	12.12

LL = liquid limit, PL = Plastic Limit, SL = Shrinkage Limit, FSI = Free Swell Index

The study was divided broadly into two segment: Study of thermal property (conductivity) of soil in its virgin state, as collected; and effect of various carbonates on the conductivity of virgin soil.

2 Lee's and Charlton's Apparatus

2.1 Construction of Apparatus

As shown in Fig. 1, Lee's & Charlton's disc apparatus consist of an aluminium disc (Lee's disc) B having diameter 10 cm and thickness 1.2 cm supported by means of four strings from a large ring on a retort stand. The experimental disc S, of the same diameter as B and thickness 1 cm, rests on it. A steam chamber C, having the same diameter as aluminium slab and thickness 6 cm, with inlet and outlet tubes, is placed above S. The bottom of the steam chamber is thick circular metal slab. Thermocouple T_1 and T_2 are inserted into holes drilled towards the base of C and into B respectively, to record their temperatures. A 5 litre round bottomed Durasil flask is placed in the heating mantle to generate steam. Soil samples of 10 cm diameter and 1 cm thickness is prepared at density 90% & 95% MDD on wet side of the OMC.

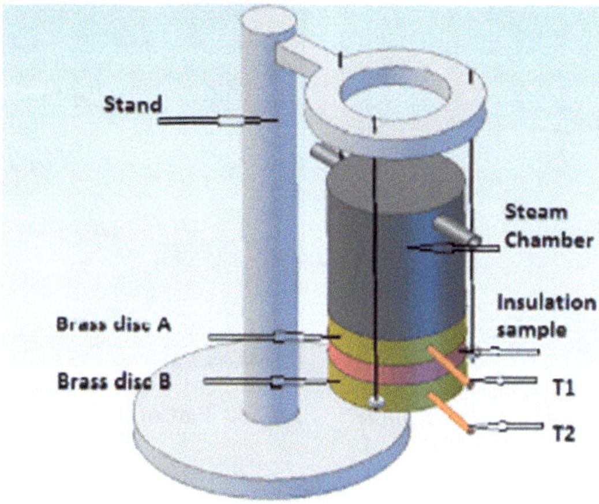

Fig. 1. Schematic diagram of Lee's and Charlton's apparatus

2.2 Sample Preparation

All samples were prepared based on Maximum Dry Density and Optimum Moisture Content. Samples were prepared at 90% and 95% of MDD with the water content being kept on the wet side of OMC. The sample size for all the experiments was 10 cm diameter and 1 cm thickness. All three soils were allowed to rest (mature) for a time duration of 24 h before using the soil for further testing. After 24 h of maturing, the soil was mixed well and shaped into cylindrical discs of 10 cm diameter and 1 cm thickness, compaction is done using small rammer to accommodate the required quantity of soil with regular measurement of thickness of the sample to get the desired uniform dimensions, as seen in Fig. 2. Samples of soil with different carbonates in different percentages, were prepared in the same manner as specified in Table 2. For all Lee's and Charlton's experiments, calculation of thermal conductivity is based on following parameters,

- Mass of Lee's Disc = 0.509 kg
- Thickness of Lee's Disc = 0.016 m
- Radius of Lee's Disc = 6 cm
- Thickness of soil sample = 1 cm
- Heat capacity of metallic Disc = 0.91 kJ/kgK

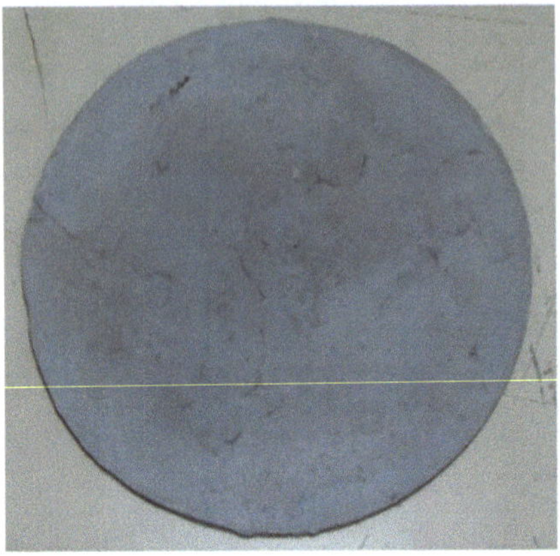

Fig. 2. Lee's and Charlton's disc sample

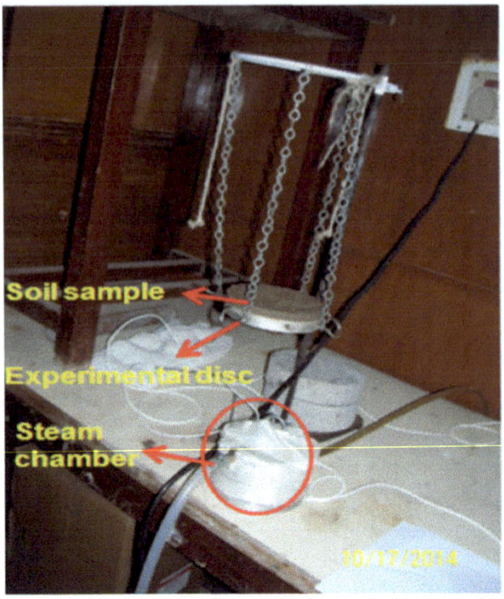

Fig. 3. Part of Lee's and Charlton's apparatus

Table 2. Composition of various samples

Sample no.	Density	Soil name	Carbonate	% Carbonate added
1	95% MDD	Kaolinite	-	-
2	90% MDD	Kaolinite	-	-
3	95% MDD	Black cotton soil	-	-
4	90% MDD	Black cotton soil	-	-
5	95% MDD	Silt	-	-
6	90% MDD	Silt	-	-
7	95% MDD	Kaolinite	Potassium carbonate	2,4,6,8,10
8	95% MDD	Kaolinite	Sodium carbonate	2,4,6,8,10
9	95% MDD	Kaolinite	Ammonium carbonate	2,4,6,8,10
10	95% MDD	Black cotton soil	Sodium carbonate	2,6,10
11	95% MDD	Silt	Sodium carbonate	2,4,6

2.3 Procedure

The mass of the aluminium disc B (Lee's Disc) was measured using a balance, whereas the diameter and thickness was measured with a vernier caliper. The arrangement of the entire setup is shown in Figs. 1, 3 and 4. The soil sample was kept between the two metal discs with the thermocouples inserted in the disc itself with the steam chamber as well as the lower metal disc. Figure 4 shows the setup for generation of steam and the connection for the steam to be delivered to the steam chamber. The thermocouples measure the temperature of the steam chamber as well as the discs at set time intervals. The steam rate is set so as to achieve a constant temperature in the steam chamber.

Once the temperature becomes constant i.e. steady the soil sample is removed between the disc and the steam chamber. The steam chamber was placed directly over the metal disc until the temperature of the lower discs rises 10 °C above the selected steady temperature. Temperature versus time reading are noted till the time that the temperature of the disc falls 10 °C below its steady state. These measurements are then used to calculate the thermal conductivity of the soil (Fig. 5).

Fig. 4. Steam passing to steam chamber

Fig. 5. Heating mantle for steam generation

2.4 Thermal Conductivity Measurement

Let θ_1 and θ_2 be the steady state temperatures recorded by the thermocouples T_1 and T_2 respectively. If d is the thickness and a is the cross sectional area of the disc S, then the quantity of heat Q conducted through the disc per second is given by

$$Q = \frac{KA(\theta_1 - \theta_2)}{d} \quad (1)$$

Where K is the thermal conductivity of the material of the disc S. Thus,

$$K = \frac{\mathrm{Qd}}{\mathrm{A}(\theta_1 - \theta_2)} \quad (2)$$

This is the working formula of the experiment. Here K is the determined by measuring A, d, θ_1 and θ_2 directly and Q indirectly,

If m is the mass, s is the specific heat, and $d\theta/dt$ is the rate of cooling at θ_2 of the brass slab B, then

$$Q = ms\left\{\frac{d\theta}{dt}\right\} \tag{3}$$

In the present method, the rate of cooling of the disc B is determined without the experimental disc S on it. So on it, so, to obtain the correct value of $d\theta/dt$ at θ_2 under the condition of the experiment, the quantity $d\theta/dt$ is multiplied by a correction factor f, due to Bedford. This factor is given by

$$f = \frac{(r + 2d_1)}{(2r + 2d_1)} \tag{4}$$

Where r is the radius and d1 is the thickness of the circular brass B
Therefore,

$$Q = ms\left\{\frac{d\theta}{dt}\right\} \times \frac{r + 2d_1}{2r + 2d_1} \tag{5}$$

Thus thermal conductivity is given by

$$K = \frac{ms\left\{\frac{d\theta}{dt}\right\} \times \frac{r+2d_1}{2r+2d_1} \times d}{A(\theta_1 - \theta_2)} \tag{6}$$

where, k- Coefficient of thermal conductivity of the sample (W/mK), A - Area of the sample in contact with the metallic disc (mm^2), d - Thickness of the sample (mm), T_2 - T_1 = Temperature difference across the sample thickness (K), m - Mass of the metallic disc (kg), s - the heat capacity of the metallic disc (kJ/kgK)d_1 - Thickness of the metallic disc (mm), r - Radius of the metallic disc (mm), $d\theta/dt$ = Rate of cooling of the metallic disc at T_2.

3 Results, Analysis and Discussion

Tables 2, 3, 4, 5 and 6 show calculations for thermal conductivity of all the three soils with carbonates (Table 7).

Figure 6 shows the thermal conductivity of Kaolinite samples prepared with potassium carbonate at different percentages. For potassium carbonate thermal conductivity increases and then decreases and then remains almost constant.

Table 3. Thermal conductivity of kaolinite with potassium carbonate

Sample no.	% MDD	Trial no.	% carbonate	Steady temperature of steam chamber θ_1 (°C)	Steady temperature of disc θ_2 (°C)	Rate of cooling $d\theta/dt$ (°C/sec)	Thermal conductivity k (W/mK)
1	95	Set 1	2	97.36	71.86	0.028	2.72×10^{-4}
2	95	Set 2	2	96.8	74.48	0.030	0.00033305
3	95	Set 1	4	96.12	72.64	0.029	0.00030604
4	95	Set 2	4	95.62	74.05	0.031	0.00035611

Table 4. Thermal conductivity of kaolinite with ammonium carbonate

Sample no.	% MDD	Trial no.	% carbonate	Steady temperature of steam chamber θ_1 (°C)	Steady temperature of disc θ_2 (°C)	Rate of cooling $d\theta/dt$ (°C/sec)	Thermal conductivity k (W/mK)
1	95	Set 1	2	95.94	64.96	0.031	0.00024795
2	95	Set 2	2	93.07	65.86	0.031	0.00028220
3	95	Set 1	4	95.35	68.2	0.033	0.00030118
4	95	Set 2	4	94.76	64.45	0.029	0.00023708

Table 5. Thermal conductivity of kaolinite with sodium carbonate

Sample no.	% MDD	Trial no.	% carbonate	Steady temperature of steam chamber θ_1 (°C)	Steady temperature of disc θ_2 (°C)	Rate of cooling $d\theta/dt$ (°C/sec)	Thermal conductivity k (W/mK)
1	95	Set 1	2	96.35	76.28	0.031	0.00038273
2	95	Set 2	2	95.63	76.77	0.033	0.00043356
3	95	Set 1	4	96.59	73.53	0.030	0.00032236
4	95	Set 2	4	95.33	75.93	0.033	0.00042149

Table 6. Thermal conductivity of black cotton with sodium carbonate

Sample no.	% MDD	Trial no.	% carbonate	Steady temperature of steam chamber θ_1 (°C)	Steady temperature of disc θ_2 (°C)	Rate of cooling $d\theta/dt$ (°C/sec)	Thermal conductivity k (W/mK)
1	95	Set 1	2	95.94	73.72	0.04	0.00044606
2	95	Set 2	2	96.07	74.42	0.036	0.00041202
3	95	Set 1	6	95.63	73.42	0.039	0.00043510
4	95	Set 2	6	95.34	74.66	0.037	0.00044333

Table 7. Thermal conductivity of silt with sodium carbonate

Sample no.	% MDD	Trial no.	% carbonate	Steady temperature of steam chamber θ_1 (°C)	Steady temperature of disc θ_2 (°C)	Rate of cooling $d\theta/dt$ (°C/sec)	Thermal conductivity k (W/mK)
1	95	Set 1	2	94.85	77.62	0.033	0.00047458
2	95	Set 2	2	94.89	79.66	0.038	0.00061824
3	95	Set 1	4	95.57	77.7	0.036	0.00049918
4	95	Set 2	4	95.5	73.59	0.032	0.00036190

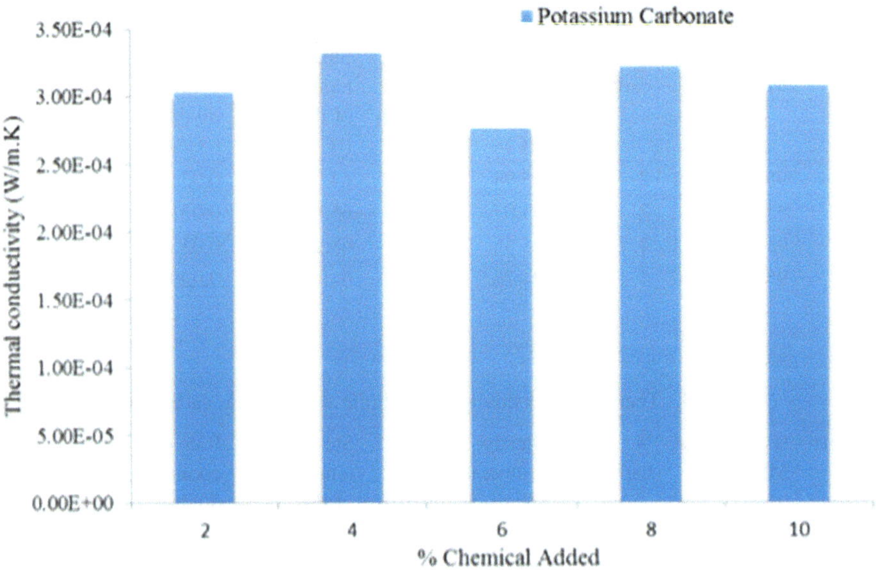

Fig. 6. Thermal conductivity of kaolinite with potassium carbonate at 95% MDD

Figure 7 shows that thermal conductivity with ammonium carbonate increases with increase in percentage of carbonate. Values of thermal conductivity at 2%, 6%, 10% are (2×10^{-4}), (3×10^{-4}), (2×10^{-4})W/m K which shows that for 6% it is the highest and then it decreases Thus after a certain amount of ammonium carbonate there is no increase on thermal conductivity of kaolinite.

Figure 8 shows thermal conductivity for black cotton soil with sodium carbonate, follows the same trend as that of thermal conductivity for kaolinite with sodium carbonate.

Figure 9 shows that thermal conductivity of silt with sodium carbonate. The value of thermal conductivity of silt with sodium carbonate at 2%, 4%, 6% are (5×10^{-4}), (4×10^{-4}), (3×10^{-4})W/m K respectively which shows that thermal conductivity of silt with sodium carbonate decreases with addition in percentage of chemical.

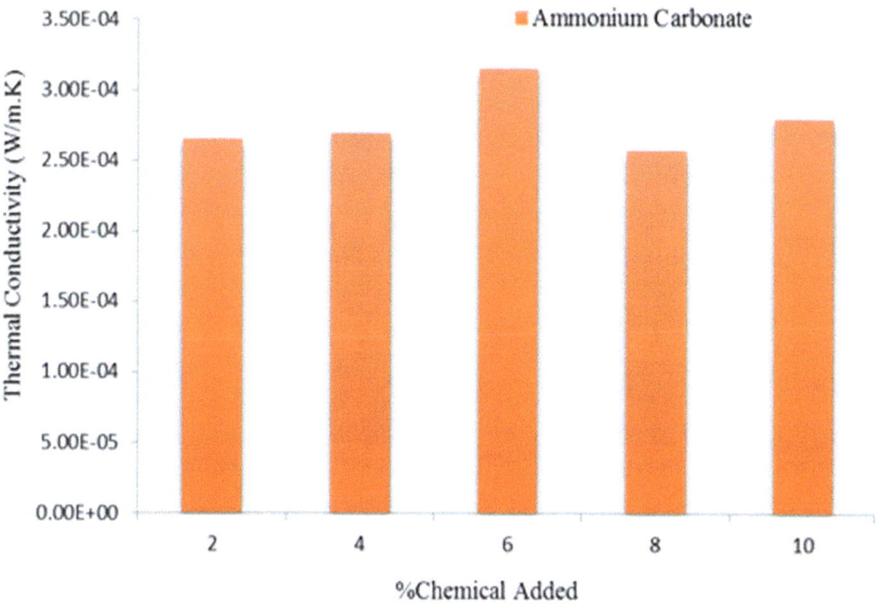

Fig. 7. Thermal conductivity of kaolinite with ammonium carbonate, at 95% MDD

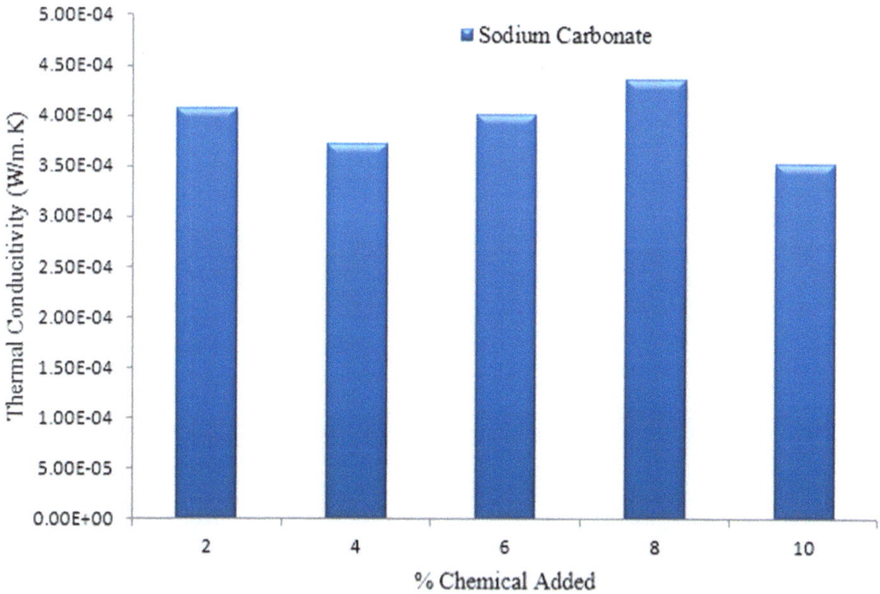

Fig. 8. Thermal conductivity of kaolinite with sodium Carbonate, at 95% MDD

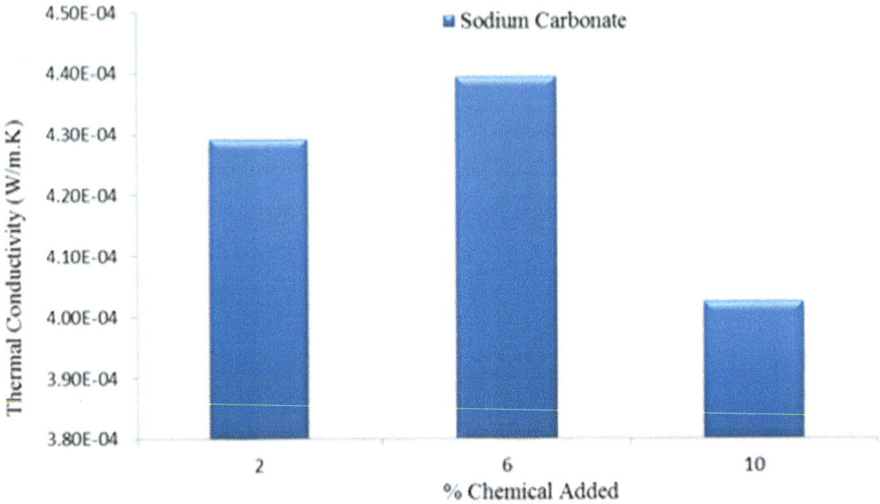

Fig. 9. Thermal conductivity of black cotton with sodium carbonate, at 95% MDD

Figure 10 compares the thermal conductivity of kaolinite, black cotton soil and silt mixed with sodium carbonate. For 2% addition of sodium carbonate thermal conductivity for kaolinite, black cotton soil, and silt is (4×10^{-4}), (4.3×10^{-4}), (5.4×10^{-4}) W/m K respectively. A similar trend of increase in the conductivity value was observed for varying percentages of sodium carbonate added to the three soils.

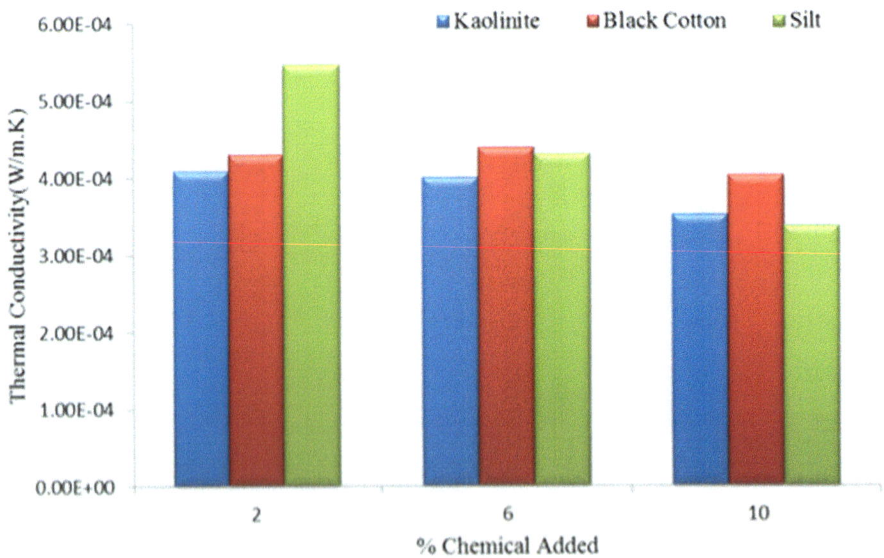

Fig. 10. Thermal conductivity of kaolinite, black cotton, silt with sodium carbonate at 95% MDD

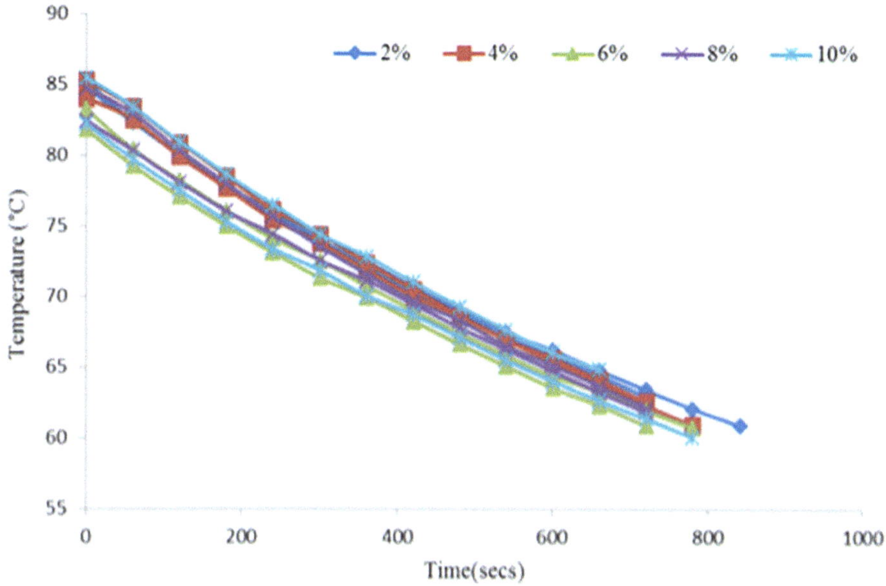

Fig. 11. Temperature v/s time (cooling rate) for kaolinite with potassium carbonate for two trials

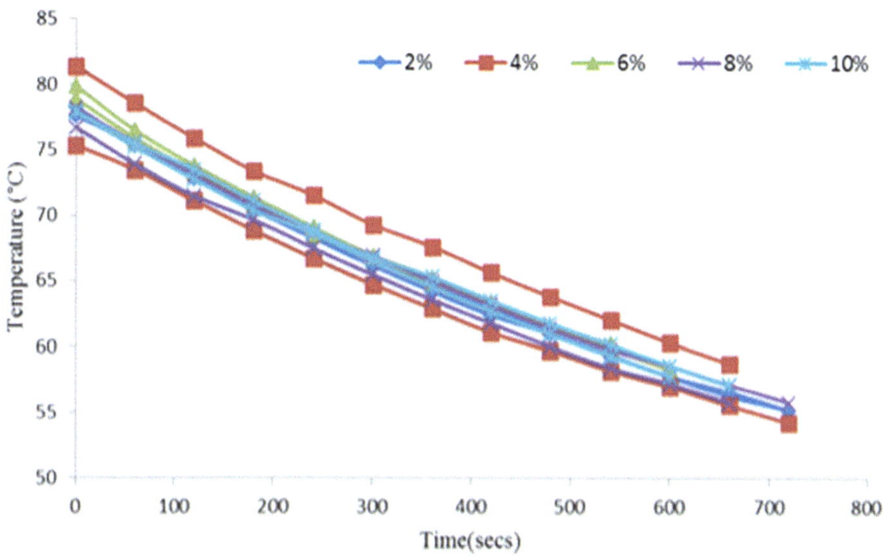

Fig. 12. Temperature v/s time (cooling rate) for kaolinite with ammonium carbonate for two trials

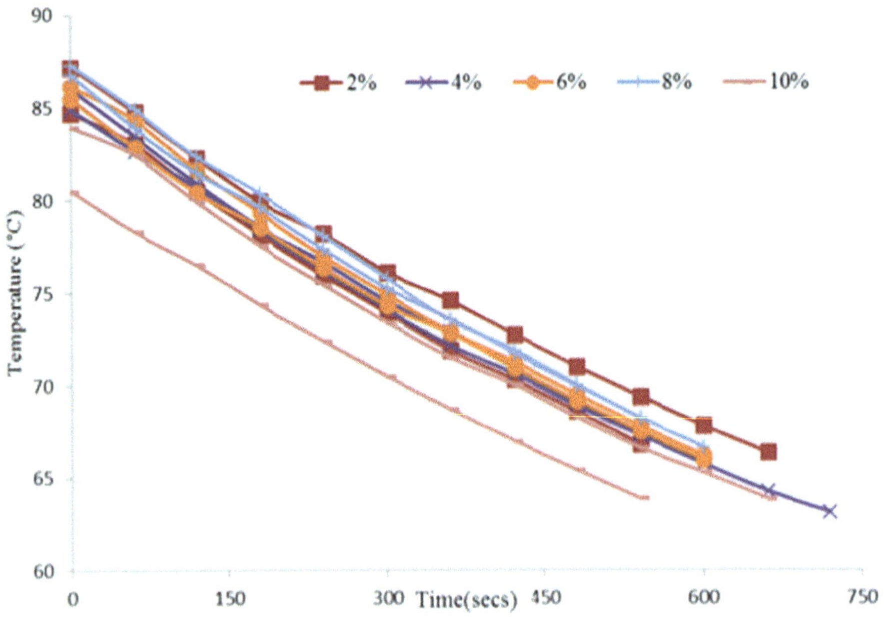

Fig. 13. Temperature v/s time (cooling rate) for kaolinite with sodium carbonate for two trials

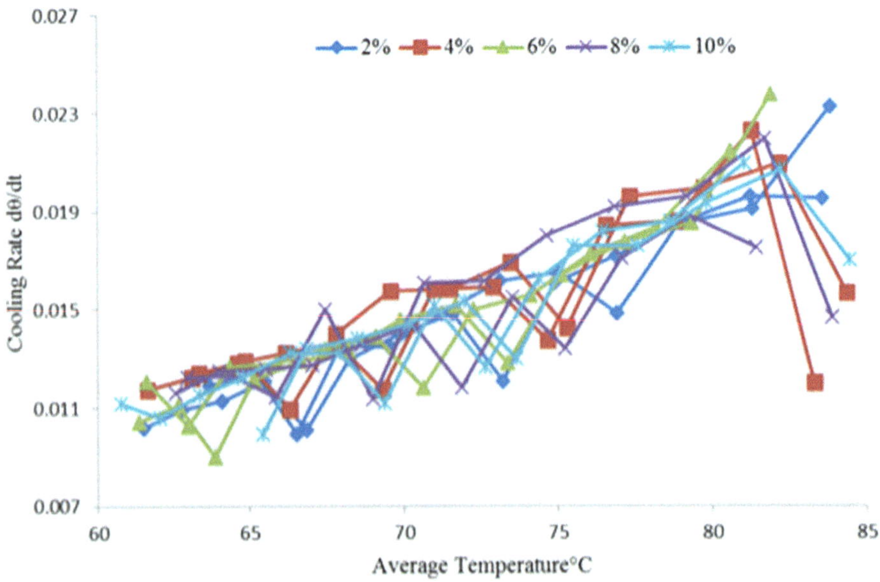

Fig. 14. Average temperature v/s cooling rate for kaolinite with potassium carbonate for two trials

Figure 10, 11 and 12 below indicates the decrease of temperature with respect to time for kaolinite with carbonates. The change of temperature with respect to time i.e. rate of cooling is calculated from the graphs. For kaolinite with sodium carbonate, the temperature range is between 80 °C to 90 °C. Similar trends were observed with black cotton soil and silt with sodium carbonate.. For other virgin soils with and without chemical mixtures temperature range is between 70 °C to 90 °C. Also, the cooling rate for virgin soils and Kaolinite with Potassium Carbonate, is between 0 to 900 secs. For all other soils with chemical mixtures cooling rate is 0 to 800 s.

The coefficient of co-relation i.e. R^2 value for polynomial trendlines are best suited when it is 1. This is in good correlation of periodic heat flow in a medium being described as propagation of thermal waves in the medium, wherein these thermal waves have characteristics similar to the characteristics of other typical waves. These plots can be further used to study waver propagation and its verification in deep soils or field conditions for study of water flux movements.

Figure 13 shows graphs of cooling rate versus average temperature for kaolinite with Potassium Carbonate. Similar graphs were drawn between other virgin soils and carbonates. The cooling rate is between 0.01 °C/sec to 0.023 °C/sec (Fig. 14).

4 Conclusions

- Thermal conductivity of kaolinite at 95% MDD is 0.0005 W/m K & at 90% mdd is 0.0003 W/m K. Similar results were observed for thermal conductivity of black cotton soil and silty soil. The observed results show that as density increases, thermal conductivity increases.
- Thermal Conductivity of virgin Kaolinite at 95% mdd is 0.0005 W/m K whereas its value drops down to 0.0002 W/m K on addition of potassium carbonate or for that matter of fact any salt used in the entire study. A comparable trend was observed for black cotton and silty soil used in the study.
- Values of thermal conductivity of Kaolinite with Ammonium Carbonate at 2%, 6%, 10% are 0.0002, 0.0003, 0.0002 W/m K respectively. It shows that for 6% addition of Ammonium Carbonate thermal conductivity of Kaolinite is highest and then it decreases after which any further increase in salt concentration fails to raise or decrease its value in any soil.
- Thermal Conductivity of silty soil with addition of 2, 4 and 6% of sodium carbonate decreases with values of 0.0005, 0.0004, 0.0003 W/m K respectively.

References

Abdulagatov, I.M., Emirov, S.N., Abdulagatov, Z.Z., Askerov S.Y: Effect of America journal, Soil Sci Soc America, pp. 8–14 (2006)

American Society of Testing Materials: ASTM D5334 “Standard Test Method for Determination of Thermal Conductivity of Soil and Soft Rock by Thermal Needle Probe Procedure”, Annual Book of ASTM Standards, vol. 0409 (2000)

Farouki, O.T.: Thermal Properties of Soils. Trans Tech Publications, New York (1986)

Aramide, F.O.: Production and characterization of porous insulating fired bricks from Ifon clay with varied sawdust admixture. J. Minerals Materials Characterization Eng. **11**, 970–975 (2012)

Flynn, D.R., Watson, T.W.: Measurements of the thermal conductivity of soils to G. Ed., vol. 1, pp. 305–343. Springer, Berlin (1969)

Bhojani, P.: Thermal conductivity of various types of soil. Maharaja Sayajirao University, Baroda, India (2014)

Salomone, L.A., Kovacs, W.D.: Thermal resistivity of soils. J. Geotech. Eng. **110**(3), 375–389 (1984)

Xiao, S.: Hydraulic conductivity investigated with Van Genuchten model (2014)

Zhenshuang, W.: Thermal properties of paraffin/SiO2 composite phase change materials. University of Finance and Economics, China

Enviro-Economic Evaluation of Dumpsite Rehabilitation Solutions in Developing Countries

A. Galaa(✉)

Geo-Institute, Housing and Building National Research Center (HBRC), 87 El-Tahrir St., Dokki, Giza 11511, Egypt
abdullah_galaa@yahoo.com

Abstract. Egypt has suffered an ill-defined solid waste management (SWM) system for decades which drove the informal sector to take control of many components within the system. As a result, uncontrolled practices prevailed leading to aggravated health, environmental, economic, and social problems. Economic losses due to poor SWM practices are magnified due to the need to deal with the environmental and health consequences. In addition, opportunities to utilize solid waste, particularly municipal and agricultural are often missed. One of the most deleterious consequences to SWM system deficiencies is uncontrolled disposal of municipal solid waste (MSW). This study presents an attempt to calculate the economic cost incurred as a result of leaving open dumps in operation and to evaluate the economic impact of four environmentally viable solutions for decommissioning open dumps. The evaluation of the four options was applied on the existing Shebin El-Kom dumpsite in Egypt. The four options proposed in the report are: Option 1: Complete removal - no waste mining (i.e. no material recovery); Option 2: Complete removal - waste mining (i.e. with material recovery); Option 3: Partial removal – capping; and Option 4: Partial removal – capping – waste mining. The comparative economic evaluation showed that Option 2 will be the most lucrative option that may return with profit and will leave the least negative impact on the public health and environment. However, other options should not be overlooked as political and administrative considerations may opt to one of them. Recommendations and required actions to refine the economic evaluation and to proceed are also presented in the paper. The same assessment methodology can be applied to other open dumping locations grown to similar scales. The key action is to integrate the closure or the decommissioning project with a wise solid waste management practice.

1 Introduction

The features of the solid waste management (SWM) system in Egypt were ill defined for a long period leading to aggravated health, environmental, economic, and social problems. Part of the inefficiencies in SWM system lay in the lack of a unified legal framework and the other part in the poor coordination between relevant public as well as private sectors. Moreover, other factors such as capital, human resource and data restrictions contribute to the problem (Zaki et al. 2013). Thus, the foundation for the decision-making process will be missing.

H. Ameen and T. Sorour (Eds.): GeoMEast 2018, SUCI, pp. 15–37, 2019.
https://doi.org/10.1007/978-3-030-01929-7_2

Uncontrolled disposal of municipal solid waste (MSW) is one of the problems that resulted from SWM system deficiencies during the past decades and is continuously growing. According to the SWEEP-NET report (Zaki 2014), 80 to 88% of the generated MSW in Egypt is open-dumped in uncontrolled dumps, 7% are disposed in landfills, and the rest is recycled and/or composted. According to the same report, out of 22 planned sanitary landfills only 7 are hitherto operational while 2 are still under construction. An urgent need exists forbetter waste management strategies to attain higher rates of waste recycling and reuse, and lower rates of disposal. At such poor recycling and reuse rates as well as insufficient landfill capacities, giant open dumps grew across the country, which poses serious threat to the environment and public health.

This report focuses on the current situation of uncontrolled open dumps in Egypt and the possible solutions to reduce the associated negative impacts. While implementing such solutions may only solves temporary and local issues, holistic changes to the SWM system should be considered and executed to ensure the sustainability of the implemented solutions.

1.1 Impacts of Open Dumps

As an open dump is approached, often randomly spread uncovered waste; scattered spontaneous fires; vermin, dogs, birds and other vectors; leachate seeping through the bottom of piles, and scavenging at site is noticed. Such features are the common characteristics of uncontrolled open dumps. However, the impact of open dumping extends to beyond unpleasant visuals to serious risk to the environment and public health as will be explained in the following sections. The training module of the United Nations Environment Program (UNEP) is consulted in the following sections addressing the different impacts.

1.1.1 Environmental Impacts

1. Contamination of groundwater and surface water due to the flow of leachate through waste and/or soil. It has also been noticed that at many locations in Egypt, solid waste is dumped by the banks of waterways or on the shores of water bodies.
2. Various contaminants (especially heavy metals) can contaminate soils beneath and beside the dump site hindering the possible future use of the site after removal of waste.
3. The emission of gases such carbon dioxide and methane due to biodegradation of organic waste contributes to the greenhouse effect.
4. Spontaneous fires (self-ignition) burns solid waste, including plastics, releasing smoke and gaseous contaminants into the air. The smoke contains particulates, carbon monoxide, carbon dioxide and other contaminant gases, including low levels of dioxins, all of which represent a serious risk to health. Odour from the generation of hydrogen sulphide may also be problematic.

5. Fauna in and around dump sites may be impacted, either by direct consumption of the solid waste, or by consumption of contaminated plants and/or animals, or as a result of leachate pollution to groundwater and surface water.
6. Plants near open dump sites can be impacted directly by the waste, dust or smoke from burning.

1.1.2 Impacts on Public Health

1. Rising smoke from dumps due to uncontrolled fires may cause chest diseases, respiratory disorders, and potentially heart diseases and cancers. Several dumpsites in Egypt are located close to residential areas which elevates such risks.
2. Polluted water and soil can directly or indirectly harm water users.
3. Infections can spread due to direct contact of scavengers with waste. In addition, vectors such as animals, rodents, birds and flying insects.
4. Waste can also slide underneath loads from moving on-site machinery causing fatal accidents.

1.2 Current Situation in Egypt

An attempt to characterize the present SWM system is presented by Egypt's SWEEP-NET country report (Zaki et al. 2013). As part of describing the system, waste disposal sites were categorized as uncontrolled dumps, controlled (public) dumps and sanitary landfills. While such classification may describe disposal sites from an administrative point of view, the reality is quite different.

Little or no controversy rises when describing sanitary landfills. The distinctive features between controlled and uncontrolled (open) dumps are detailed by the Training Module of the United Nations Environment Program (UNEP 2005). The current situation shows that uncontrolled dumps grow in random sites selected by the public with no intervention from the municipalities. Controlled dumps may only differ in the fact that municipalities officially allocate a specific site for MSW disposal with no or meagre measures to control any impact waste may impose. The only difference could be a fence. Waste disposal practices in both controlled and uncontrolled dumps in Egypt are the same. Uncontrolled dumps vary from small (few hundred square meters) to large size (thousands square meters) dump sites, and controlled dumps are mostly larger in size, while practices remain the same. At various locations, uncontrolled and controlled dumpsites both grow to occupy a gigantic footprint (several feddans/acres). El-Montazah and Abees dumpsites in Alexandria; and El-Wafa Wa El-Amal dumpsite in Cairo are good examples on that. No daily covers, no leachate/gas management and waste pickers prevail when it comes to salvaging recyclables. Based on UNEP's description both controlled and uncontrolled dumps in Egypt can be classified as Uncontrolled Open Dumps.

2 Possible Gains from Dumpsite Rehabilitation

The previous sections addressed the possible harmful impacts of uncontrolled dumps and the associated practices on the public health and safety as well as on the environment. The required remedial measures range from decommissioning of the dump, whether by removal or capping, to the upgrade to a more controlled waste disposal facility. Choice of the most appropriate remedial measure depends on environmental as well as economic factors. In this section, the sources of possible economic return will be presented.

2.1 Return on Environment and Public Health

Despite the damage already caused by the dump, the cost of not closing the dump will exceed the cost of closing it and constructing a properly engineered disposal facility, mainly due the impact on environment and public health. Such costs appear on the long-term and may not appear as a number in a feasibility study as, sometimes, the damages to environment and public health are immeasurable in monetary value. Containing the source of contamination will definitely stop ongoing pollution, limit the extent of polluted zone and restore reversible damage caused by the dump.

2.2 Financial Return

2.2.1 Waste Value

Both incoming and accumulated (aged) waste at a dump are of economic value. Incoming fresh waste can be diverted to sorting and recycling plants, composting facilities, and waste-to-energy plants. The possible returns from accumulated waste are presented in this paper. If the dump is to be totally or partially removed, the removal process can be turned to mining and processing activities instead of just transporting the waste to new locations. Accumulated waste can be mined for the following:

1. Compost: the organic fraction in aged waste has undergone considerable aerobic and/or anaerobic digestion and may need only minor processes to reach an end product, given that the waste has not undergone severe incineration.
2. Recyclables: accumulated waste is already scavenged and may contain a fraction of recyclables significantly less than fresh waste. However, there is always a fraction of recyclables left in the waste that could be of economic value such as scrap electronics and aluminium (Hogland et al. 2011).
3. Rejects-to-energy: The Energy Rich Fraction (ERF) of the non-biodegradable and non-recyclable materials in accumulated waste can be turned into refuse-derived-fuel (RDF). Combustibles obtained from aged waste have lower moisture content and higher calorific value than those obtained from fresh waste as proven by Chen et al. (2007).

2.2.2 Land Value

Soil underlying open dumps experience serious contamination that may hinder the reuse of the land for various activities, particularly agricultural purposes. Treatment of soil contamination is a costly, complicated and lengthy process. The good side is having several suitable uses for the affected land. The location and post-closure use of land defines the potential value of the salvaged land through removal or capping of the open dump. There have been examples in Egypt where a capped dump has been used for recreational purposes after decommissioning (closure). For example, EL-Darassa dumpsite was capped and turned into what is currently known as Al-Azhar Park giving live example of turning a dumpsite to a profit generating project when the location is lucrative. Al-Fustat garden is another example on a smaller scale. If the location is not as lucrative as in the previous cases, the dump could be partially or totally removed and land can be utilized in a more efficient waste management system. For example, transfer stations; sorting, composting, and/or recycling plants could be built on the cleared section of the salvaged land. This was the solution adopted for Al-Montazah dumpsite in Alexandria.

3 Dump Decommissioning Project Components

The dump decommissioning project has three interdependent components, namely, an engineering component pertaining to excavation/reshaping/capping of the dump; a SWM component including the integration of the dump in an efficient SWM and material recovery; and the post-closure land use component. The SWM and the land use components were discussed in the previous section on the economic return while the engineering component will be detailed in the following section. However, each component cannot be considered separately as several combinations of the different components may be suitable.

3.1 Engineering Solutions

Several engineering solutions could be served at the same site depending on the ultimate goal behind decommissioning and on the cost of the best combination of the project components. Engineering solution for decommissioning a dump can fall under four main methods:

1. Decommissioning by completely removing the waste and remediating the site,
2. Decommissioning by shaping and capping the waste,
3. Combining 1 and 2, and
4. Decommissioning the uncontrolled practices by upgrading the site to a sanitary landfill

The first three methods start by preventing dumping at the current site and allocating an alternative site for waste disposal to accommodate all or a fraction of the amount of waste at the dump as well as the incoming waste. Upgrading to a sanitary landfill may not be suitable for most of the major uncontrolled dumpsites in Egypt due to the inappropriateness of their locations for waste disposal in the first place. Therefore, the following sections will discuss the first three methods.

3.2 Complete Removal and Site Remediation

There are two major components for this solution that may seem conventional, excavation and transportation. However, MSW is not like conventional excavation debris and require special precautions especially during excavation. Methane is explosive when exposed to heat at concentrations between 4.4% to 15% (UNEP 2005). Therefore, the use of spark-less equipment and machinery may be necessary. If such equipment is not available, methane levels in the air should be continuously monitored and work should be halted until levels are back to below the explosive limits of methane.

Transportation of the excavated waste for disposal in an alternative location or for processing at a recycling or composting facility is the other major component of complete removal solution. Therefore, waste destination should be prudently selected to minimize the cost of transportation. The best practice is to process the material onsite and transport only a small fraction to the closest sanitary landfill.

Leachate and gas migration into the soil and water underneath the dump causes a serious environmental challenge. Remediation of the contaminated soil, groundwater, surface water, and air is required upon the removal of the dump. Remediation options range from removing the source of contamination or containing the contamination to the cleanup of contaminated zones. The suitable option depends of the foreseeable use of the salvaged land and on the scale of contamination caused by the dump. Site assessment and investigation is required to assess the scale of the problem and select the suitable option. Soil and groundwater should be tested for the concentration of the different contaminants commonly available in MSW leachcate.

3.3 Shaping and Capping the Dump

This alternative is preferred when removing the waste is not feasible or practical. This alternative contains the waste in-place and minimizes its impact on the surrounding environment (air, soil, and water). This happens via capping the top surface of the dump with a low permeability cover or a barrier while installing gas and leachate collection systems to control the migration of such contaminants. According to Solid Waste Management Privatization Procedural Manual (SWMPP) and UNEP (2005), the main elements of this solution are:

- **Shaping (cut and fill) and compaction** of waste to the required shape with stable slopes and graded surfaces.
- **Final cover (cap)** to minimize the percolation of water to covered waste, reduce gas migration, isolate the waste from vermin and animals, minimize the escape of odor, and to support vegetation for landscaping purposes. The cover is composed of a gas collection soil layer on the top of waste, overlain by compacted low permeability soil or alternatively geotextile and geomembrane (liner) for preventing water flow, overlain by uncompacted topsoil for vegetation and landscaping purposes.
- **Drainage control system** through providing the adequate slopes and water pathways to convey water to particular spots where it can be collected and disposed. Slopes and materials should also be designed to avoid surface erosion.

- **Leachate management system** consists of a series of pipes, drains, ditches and/or canals located to intercept leachate, collect it, and transfer it to a leachate pond where it can be treated. It is also possible to provide containment elements such as cutoff walls and interceptor trenches to limit the migration of leachate.
- **Gas management system** to vent or flare the accumulated gases (particularly methane) and ensure that it does not build up pressure or migrate laterally to adjacent structures through easiest paths within the soil. Gas is extracted or vented through venting pipes or wells installed at regular spacing. Landfill gas can be collected by either a passive or an active collection system. Passive gas collection systems use existing variations in landfill pressure and gas concentrations to vent landfill gas into the atmosphere or a control system; while active systems include vacuums or pumps to move gas out of the landfill and piping that connects the collection wells to the vacuum (Qian et al. 2002).
- **Post-closure monitoring wells** where upgradient and downgradient wells from the closed dumpsite should be installed to test for any gas or water contamination. Periodic testing of these wells is recommended until waste stabilizes.

3.4 Combining Removal and Capping (Hybrid)

It is possible to partially remove the waste or only clear part of the land for particular use while keep the rest of the dump in place and cap it. This option depends mainly on the planned post-closure land use and on the cost of the different components of the project. For instance, the cost of transporting excavated waste to an alternative site could be high if the closest alternative location is far from the dump or if the available area of the dump is not enough to accommodate the residing waste after shaping and grading the dump to satisfy slope stability and erosion control considerations. In addition, waste management facilities (e.g. composting plants or transfer station) could be built on a small part of the land.

4 Economic Evaluation

In the previous section, the components of decommissioning an uncontrolled dump were presented and it is obvious that practicing any combination of the components will cost money. In this section the main elements of a feasibility analysis will be presented.

Economic evaluation of decommissioning a dump is, as most environmental projects, a complex procedure and it incorporates more components than those explained earlier. The project components presented in the previous section define what is called "internal costs" or "financial costs". However, other costs referred to as "external costs" or "externalities" come into the equation in Economic evaluations. According to a report published by the Directorate-General for the Environment, European Commission (DGE-EU), the external costs are defined as (COWI 2000):

> *"The costs which arise when the social or economic activities of one group of people have an impact on another, and when the first group fail to fully account for their impact."*

The following sections give more details on the expected internal and external costs to be identified in an economic evaluation for decommissioning a dumpsite.

4.1 Evaluation of Financial (Internal) Costs

To study the feasibility of any combination of the project components, it is important to outline the expected costs and returns for each of the project components. Table 1 shows a cost-return matrix to facilitate the decision making if any of the project components is part of the options under consideration.

Table 1. Financial evaluation matrix for the different potential project components

Project component		Costs	Expected return
1. Engineering component	A. Removal	• Waste excavation and handling • Transportation • Landfill occupation	
	B. Capping	• Waste cut and fill (shaping) • Compaction • Cover layers • Leachate and gas management systems • Drainage system	
	C. Hybrid	Percentage of A plus Percentage of B	
2. SWM component	A. Transfer station	• Waste excavation and handling • Transportation • Containers and trucks • Station construction	
	B. Mining and processing	• Investors pay all costs for composting and/or RDF production (excavation, equipment rental, sieves, transportation, further processing, packaging, etc.)	• Get-fee per ton of compost • Get-fee per ton of RDF • Get-fee per ton of recyclables (if any)
3. Post-closure land use component	A. After Removal	• Land remediation • Groundwater and gas monitoring	• Sale value, or • Rental amount, or • Concession value
	B. After Capping	• Vegetation and landscape • Groundwater and gas monitoring	• Rental amount, or • Concession value
	C. Hybrid	Percentage of A plus Percentage of B	• Rental amount, or • Concession value

4.1.1 Assumptions

There some assumptions to be made in order to clarify the value of the waste and to facilitate the decision making procedure:

- Any combination should account for the cost of finding an alternative location for disposal.
- The project could be funded by the municipalities (or other relevant authorities) or investors could be called to bid for land concession (e.g. BOT).
- A fee could be collected in return for utilizing the waste in composting, recycling, and waste-to energy activities. This is a fee per unit weight of the recovered or processed material.
- Aged waste has already underdone aerobic and anaerobic reaction and may need only minor processing to be turned into compost.

Examples of the different combinations from the matrix above can be as follows:

- **Option 1:** complete removal of the dump without SWM or mining (material recovery). In this case, there will be the cost of removal, land remediation, and preparing an alternative location while the return will come from the post-closure land use.
- **Option 2:** complete removal with SWM solutions/waste mining. In this case, there will be the cost of removal, land remediation, and preparing an alternative location while the return will come from the post-closure land use in addition to the sale of recovered material. This combination will offer greater return than the first option while longer time is needed for material recovery and processing.
- **Option 3:** Partial removal and covering the remaining waste. In this case, there will be the costs of removal, capping (as in Table 1), and preparing an alternative location while the return will come from the limited uses of the land (mostly recreational).
- **Option 4:** Partial removal and waste mining and covering the remaining waste. In this case, there will be the costs of removal, capping (as in Table 1), and preparing an alternative location while the return will come from the limited uses of the land (mostly recreational) and from material recovery.

There are several combinations that could be considered and compared for the same project. The decision will be taken based on the combination that will provide the best economic and socio-environmental values.

4.1.2 Cost of Alternative Location

An alternative location could be a landfill, recycling/composting/RDF plant, and/or a transfer station. An existing landfill within an adequate distance could be an option. However, the lifetime of the landfill will be reduced by the addition incoming waste, which is translated into cost. The best practice is to implement a waste management strategy (if not in place) to ensure that any investment made in finding an alternative destination for the waste pays back a significant amount of the investment and potentially returns profits.

4.2 Evaluation of External Costs

While quantifying or monetizing all the external costs, such as cost of damage to public health and the environment, from waste disposal is difficult, COWI (2000) followed a methodology referred to as the "Impact Pathway Method (IMP)" to quantify the damage caused by a dumpsite. In this methodology, the report studied the externalities ensuing waste incineration and waste disposal separately, as for quantifying the impact of each on receptors is different. The impacts of incineration and waste disposal are quantified by identifying the damages.

In the preceding paragraph three terms were introduced, **impact**, **receptor** and **damage**. For waste incineration and disposal the identified **impacts**, according to COWI (2000), could be summarized as:

- Emissions to air: such as particulates, dioxins, nitrogen oxides, heavy metals, acid gases, methane, CO_2, and volatile organic compounds (VOC).
- Emissions to soil and water: such as dioxins, heavy metals, suspended solids, and leachate.
- Disamenity: such as noise, dust, odours, and visual pollution.
- Residual solid waste
- Risk of accidents

Receptors include humans, fauna, flora, and buildings. However, the practice is to study the adverse effects on humans.

To quantify the cost of exposing receptors to impacts, the following were identified as the expected **damages**:

- Human health effects: mortality and morbidity
- Lower agricultural yield
- Forest die-back or devegetation
- Damage to buildings
- Climate change
- Effects on the ecosystem.

The sequence of applying the IPM is presented in Fig. 1. In order to estimate the external costs in an economic evaluation, two key figures are essential and should be obtained for each pollutant (see Impacts): Emission factors (measured in kg per tonne of waste) and Unit costs (measured as cost per kg emission).

The following section will present a real case in Egypt where practical solutions and combinations from the matrix above are compared and a best solution for the decommissioning of the dump is recommended.

5 Case Study: Shebin El-Kom Open Dump

The location of the dump was allocated by municipality with no record of when dumping started at the location. It is considered a public dump and all practices should classify it as an uncontrolled dump as will be shown later. The dump serves mainly the major two districts (Eastern and western districts) of the city of Shebin El-Kom and

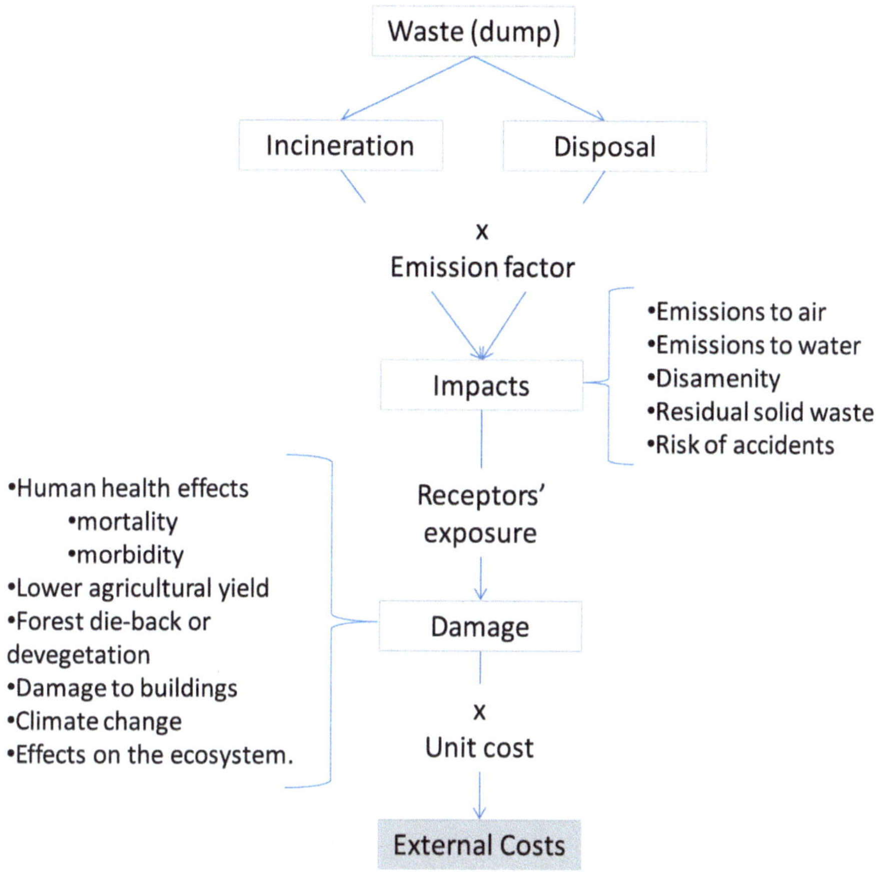

Fig. 1. Impact Pathway Method from waste disposal to cost

other villages close by. The population of the Eastern and Western districts is 195,300 capita according the official estimates by the Monofia Governorate in 2014. According the Environmental Characterization published by The Egyptian Environmental Affairs Agency (EEAA) in 2008, the Shebin El-Kom dump received 160 tons/day in 2008. Based on this tonnage and an estimated annual growth of population of 2.5%, the estimated tonnage of waste received at the dump in 2018 is approximately 205 tons/day.

5.1 Current Situation

The dump is located in the Western District surrounded by agricultural land and in the vicinity of the residential areas (less than 100 m away) as shown in Fig. 2. The latitude and longitude an arbitrary point in the center of the dump are 30.543395° and 30.999164°, respectively, covering an area of approximately 3.7 Feddans (3.84 acres). Elevations of the natural ground at the base of the dump ranges from 12 to 14 m above sea level (ASL) and at the highest point of the dump around 39 m ASL, indicating that the dump reaches a height of around 26 m. The estimated volume of the dump, according to measurements done at the time of the visit, is around 240,000 m^3 and estimated tonnage of waste is 192,000 tons based on a density of 0.8 t/m^3.

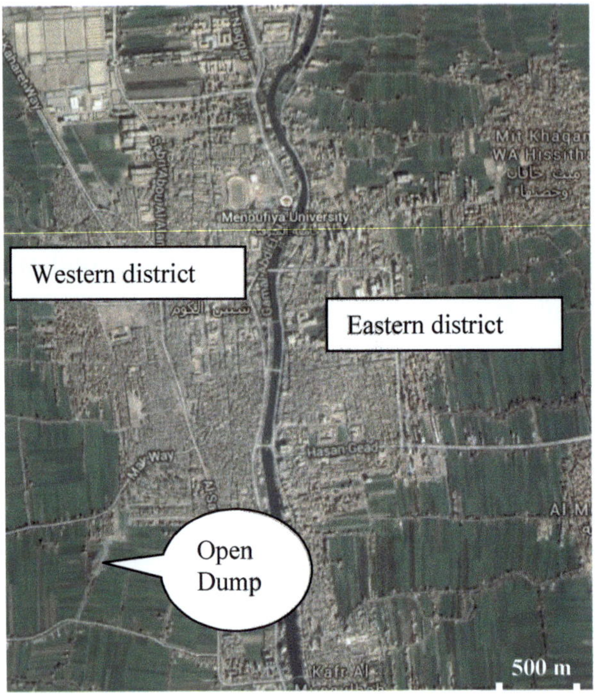

Fig. 2. Eastern and Western districts of Shebin El-Kom city and the location of the dump

Approaching the site reveals overwhelming smoke and musty odour emitted from a mountain of waste as shown in Fig. 3. By walking around the dump, one can see numerous spots of self-ignited fires. However, the exposed aged waste showed that only the surface is burnt to ashes while the inside is not. Leachate was also seen leaking through the bottom of the dump at various locations as shown in Fig. 4. In addition, adjacent land north of the dump was submerged by leachate showing signs of dead vegetation as shown in Fig. 5.

There have been various activities taking place at the dump. Municipality trucks take an up-ramp to the top of the dump and unload waste where waste pickers grub for recyclables. Excavators and loaders were excavating at the bottom of the dump and load trucks to dispose in Al-Sadat landfill 35 km away from the dump, which resulted in an exposed vertical waste cliff posing risks of imminent sliding. The excavation effort is an attempt from the municipality to reduce the size of the dump. However, incoming waste are still disposed at the same location. There is another attempt to reduce the size of the dump through the presence of a compost trommel screen onsite.

Fig. 3. Smoke rising from the dump as approached from the north side

Fig. 4. Leachate seeping through the toe of the dump

Fig. 5. Dead vegetation in leachate-soaked land (north of the dump)

The screen is capable of producing 50 tons of compost per day as claimed by screen operator. In return, the compost producer pays around LE 8,000/month to the municipality, which translate to approximately 8 LE per ton of compost. It is important to note that no daily cover is added, no recording of incoming waste, and all other practices are indicators of a major uncontrolled dump.

5.2 Evaluation of Decommissioning Options

In this section, the expected internal costs and returns at this particular location will be presented various options for decommissioning the dump will be detailed and compared. However, data required to perform the necessary analysis to valuate external costs are not available at this stage of the study. Therefore, the report will only refer to damage incurred before decommissioning and or damage alleviated in the option under consideration. This evaluation will lead to a recommendation of the most beneficial option with respect to financial, economic, and environmental aspects. Costs and revenues estimates are based on unit rates collected at the time of the study prior to the devaluation of the local Egyptian currency that took place in November, 2016.

5.3 Estimation of Internal Costs and Return

Estimation of costs and return is based on assuming certain unit prices based on experience with similar projects in Egypt and market surveys. As elaborated earlier, the cost of an alternative location is a fixed cost in all options under consideration. Post-closure groundwater monitoring and soil remediation is component that may vary depending on the decommissioning option under consideration and post-closure land use. On the other hand, the return from selling or renting the salvaged land should be fixed if the waste is totally removed. The following sections will present the expected fixed costs and returns before analyzing the cost and return for each of the options under consideration.

5.3.1 Cost of Alternative Location (Transfer Station)

While selecting an alternative location, care should be given to adding value. This can be achieved by changing the SWM practices from just-dumping to better utilization of waste. Out of this belief, it is suggested to allocate a location of approximately a feddan within the outskirts of the city of Shebin El-Kom as a transfer station. From the transfer station, waste could be transported to Qowesna or El-Shohada mechanical and biological treatment plants, 13 and 16 km away, respectively. If the capacity of both plants cannot accommodate the incoming waste from Shebin El-Kom, waste can be transported from the transfer station to El-Sadat landfill, 35 km away from Shebin El-Kom, until a recycling and compost plant is built to accommodate the waste generate by the city. The total expected cost of the transfer station that will accommodate approximately 200 tons per day is detailed in Table 2.

Table 2. Estimated cost of an alternative transfer station in Egyptian Pounds (LE)

	Item	Quantity	Unit price (LE)	Total (LE)
1	Container with press	4	30,000	120,000
2	Truck head	2	800,000	1,600,000
3	Construction of ramps, platform, tire wash facility, sheds and utilities	1	500,000	500,000
	Total			**2,220,000**

5.3.2 Return on Land Utilization

Advancements in environmental sciences enable the after use of land for any purpose after remediation. However, using the land in agriculture or any activity related to the health and hygiene will require extensive remediation measures at high expenses. Therefore, it is recommended to prohibit the use of the land in the aforementioned activities. With less stringent cleanups or soil removal, the salvaged land could be used for residential, commercial, or recreational purposes. However, the remediation strategy and cost are subject to a full scale environmental investigation. The average price is 2,000 Egyptian Pounds (LE) per square meter for residential purposes. Land can be sold for commercial (e.g. malls) or recreational (e.g. entertainment/theme parks) purposes for the same price. Based on this estimate, the 3.7 feddans could be sold for approximately 20 to 25 Million Egyptian pounds after reducing the amount for the sake of open spaces for access, landscape, utilities and services.

It is possible to remediate the land before sale or leave this task to the buyer. However, in the options discussed below, the cost of remediation will be added to the cost of the project. The following sections will elaborate on the economic evaluation of different options for combining the components of the decommissioning project.

5.4 Decommissioning Options for the Shebin El-Kom Open Dump

5.4.1 *Option 1: Complete Removal - No Waste Mining*

In this option, it is assumed that:

- Dumping will be prevented before starting the decommissioning project.
- Waste will be excavated and transported to El-Sadat landfill 35 km away from the City of Shebin El-Kom.
- The cost of excavating and transporting the waste is 70 LE/ton.
- Quantities are those estimated on the day of field investigation (January 31, 2015) totalling 192,000 tons (approximately 200,000 tons).
- Complete removal takes 6 to 7 months.
- Remedial measures and continuous monitoring of ground water and gases will be required after removal. Cost of remediation is uncertain as data has to be collected to assess the level of degradation. However, the cost of remediation is estimated based on complete the removal and replacement of soil within the top 2 m.
- Landfill occupation cost based on the reduction of the service life of El-Sadat landfill could not be determined based on the available data.

The economic evaluation for Option 1 is detailed in Table 3.

5.4.2 *Option 2: Complete Removal – Waste Mining*

In this option, it is assumed that:

- Dumping will be prevented before starting the decommissioning project.
- Quantities are those estimated on the day of field investigation (January 31, 2015), totalling 192,000 tons (approximately 200,000 tons).
- Composting can be done on-site via portable trommel screens.
- Produced compost can reach one third the available waste by weight (around 70,000 tons).
- Each screen can produce 50 tons of compost per day.
- Compost producer may pay to authorities 20 LE per ton produced.
- The site can accommodate 10 screens running simultaneously.
- Material recovery will delay it to 12 months if 10 screens are deployed on-site.
- The reject (around 130,000 tons) will be transported to El-Sadat landfill 35 km away from the City of Shebin El-Kom.
- The cost of transporting the waste is 50 LE/ton.
- Marginal RDF amounts are expected and will discourage any investment.
- Remedial measures and continuous monitoring of ground water and gases will be required after removal. Cost of remediation is uncertain as data has to be collected to assess the level of degradation. However, the cost of remediation is estimated based on complete the removal and replacement of soil within the top 2 m.
- Landfill occupation cost is reduced in comparison to Option 1.

The economic evaluation for Option 2 is detailed in Table 4.

Table 3. Economic evaluation for Option 1

	Status	From/incurred costs, LE	To/return, LE
Impact	Waste volume at site (tons)	200,000	Zero
	Occupied land (feddans)	3.7	Zero
	Maximum estimated gas emissions (megagrams/year)	3500	Zero
	Estimated leachate generation (m^3/year)	17,000	Zero
	Incineration	Exists	Eliminated
	Risk of accidents	High	Eliminated
	Disamenity	Exists	Eliminated
Evaluation of externalities	Cost of mortality	Exists	Reduced to eliminated*
	Cost of morbidity	Exists	Reduced to eliminated*
	Cost of lowered agricultural yield	Exists	Eliminated
	Cost of devegetation	Exists	Eliminated
	Cost of damage to buildings	Exists	Eliminated
	Cost due to climate change	Exists	Eliminated
	Cost of ecosystem degradation	Exists	Reduced to eliminated*
Estimation of internal costs and return	Removal and transportation of 200,000 ton (70 LE x 200,000)	14,000,000	
	Capping and LMS	-	-
	Construction of transfer station	2,220,000	
	Remediation and monitoring**	2,000,000	
	Material recovery	-	-
	Sale or rental value of land	-	20,000,000
Net	**1,780,000 LE + reduced to eliminated externalities**		

*Depends on the level of remediation applied to site

**Cost of remediation is uncertain as data has to be collected to assess the level of degradation

5.4.3 *Option 3: Partial Removal – Capping*

In this option, it is assumed that:

- The maximum waste slopes after shaping is 1H:3 V and the minimum width at the top of the shaped waste is 6 m, will lead to a maximum capped volume of 123,000 m^3 (i.e. 99,000 tons). Figure 6 shows how would the capped surface look.
- Around 100,000 tons will be excavated and transported to El-Sadat landfill 35 km away from the City of Shebin El-Kom.
- The cost of excavating and transporting the waste is 70 LE/ton.
- Capping will cost 150 LE/m2, including compaction, gas collection layers, liners, soils and vegetation.

Table 4. Economic evaluation for Option 2

	Status	From/incurred costs, LE	To/return, LE
Impact	Waste volume at site (tons)	200,000	Zero
	Occupied land (feddans)	3.7	Zero
	Maximum estimated gas emissions (megagrams/year)	3500	Zero
	Estimated leachate generation (m^3/year)	17,000	Zero
	Incineration	Exists	Eliminated
	Risk of accidents	High	Eliminated
	Disamenity	Exists	Eliminated
Evaluation of externalities	Cost of mortality	Exists	Reduced to eliminated*
	Cost of morbidity	Exists	Reduced to eliminated*
	Cost of lowered agricultural yield	Exists	Eliminated
	Cost of devegetation	Exists	Eliminated
	Cost of damage to buildings	Exists	Eliminated
	Cost due to climate change	Exists	Eliminated
	Cost of ecosystem degradation	Exists	Reduced to eliminated*
Estimation of internal costs and return	Removal and transportation of 130,000 ton (50 LE x 130,000)	6,500,000	
	Capping and LMS	-	-
	Construction of transfer station	2,220,000	
	Remediation and monitoring**	2,000,000	
	Material recovery (20 LE x 70,000)	-	1,400,000
	Sale or rental value of land	-	20,000,000
Net	**9,280,000 LE + reduced to eliminated externalities**		

*Depends on the level of remediation applied to site
**Cost of remediation is uncertain as data has to be collected to assess the level of degradation

- Estimated cost of leachate management system (LMS) including the installation of horizontal drains is around 300,000 LE.
- Dumping will be prevented before starting the decommissioning project.
- Partial removal takes 3 months, shaping and covering takes 4 months.
- Mild remedial measures will be required after removal. However, continuous monitoring of ground water and gases is required. More wells are expected than in Options 1 and 2 due to the continuing presence of waste under the cap.
- Landfill occupation cost is reduced in comparison to Options 1 and 2.
- Land use in this option is limited and mostly for recreational purposes.
- Land may not be sold and possibly a concession will be given to an investor for a limited period (possibly 10 to 20 years)

The economic evaluation for Option 3 is detailed in Table 5.

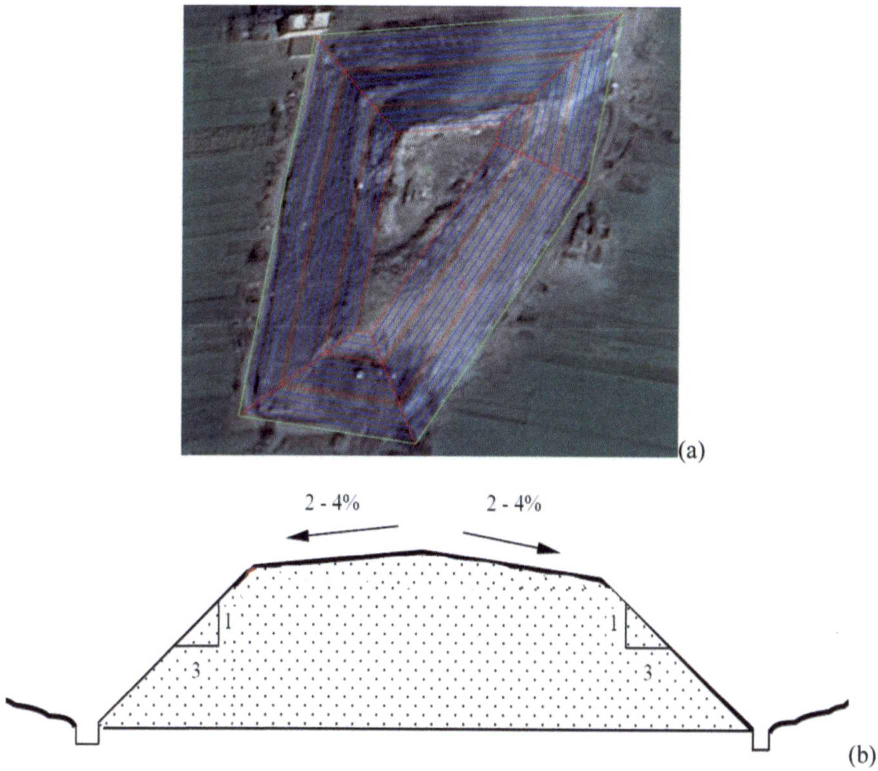

Fig. 6. Examples of (a) surface contours in plan, and (b) a section in the waste, for the capped dump

5.4.4 *Option 4: Partial Removal – Waste Mining – Capping*

In this option, it is assumed that:

- The maximum waste slopes after shaping is 1H:3 V and the minimum width at the top of the shaped waste is 6 m, will lead to a maximum capped volume of 123,000 m^3 (i.e. 99,000 tons). Figure 6 shows how would the capped surface look.
- Dumping will be prevented before starting the decommissioning project.
- Composting can be done on-site via portable trommel screens.
- Produced compost can reach one third the removed waste by weight (around 33,000 tons).
- Each screen can produce 50 tons of compost per day.
- Compost producer may pay to authorities 20 LE per ton produced.
- The site can accommodate 5 screens running simultaneously.
- Material recovery will delay it to 5 months if 5 screens are deployed on-site.
- The reject (around 67,000 tons) will be transported to El-Sadat landfill 35 km away from the City of Shebin El-Kom.
- The cost of transporting the waste is 50 LE/ton.
- Marginal RDF amounts are expected and will discourage any investment.

Table 5. Economic evaluation for Option 3

	Status	From/incurred costs, LE	To/return, LE
Impact	Waste volume at site (tons)	200,000	99,000
	Occupied land (feddans)	3.7	3.7
	Maximum estimated gas emissions (megagrams/year)	3,500	1,700
	Estimated leachate generation (m^3/year)	17,000	8,500
	Incineration	Exists	Eliminated
	Risk of accidents	High	Eliminated
	Disamenity	Exists	Reduced
Evaluation of externalities	Cost of mortality	Exists	Reduced*
	Cost of morbidity	Exists	Reduced*
	Cost of lowered agricultural yield	Exists	Reduced*
	Cost of devegetation	Exists	Reduced*
	Cost of damage to buildings	Exists	Eliminated
	Cost due to climate change	Exists	Reduced*
	Cost of ecosystem degradation	Exists	Reduced*
Estimation of internal costs and return	Removal and transportation of 100,000 ton (70 LE x 100,000)	7,000,000	
	Capping and LMS (150 LE x 4200 x 3.7 + 300,000 LE)	2,630,000	
	Construction of transfer station	2,220,000	
	Remediation and monitoring**	500,000	
	Material recovery	-	-
	Sale or rental value of land	-	5,000,000
Net	**−7,350,000 LE *** + reduced externalities**		

*Depends on the level of remediation applied to site
**Cost of remediation is uncertain as data has to be collected to assess the level of degradation
***Although the project resulted in a negative net value, the land will return to the government after the concession period ends.

- Capping will cost 150 LE/m^2, including compaction, gas collection layers, liners, soils and vegetation.
- Estimated cost of leachate management system (LMS) including the installation of horizontal drains is around 300,000 LE.
- Mild remedial measures will be required after removal. However, continuous monitoring of ground water and gases is required. More wells are expected than in Options 1 and 2 due to the continuing presence of waste under the cap.
- Landfill occupation cost is reduced in comparison to Options 1, 2, and 3.
- Land use in this option is limited and mostly for recreational purposes.
- Land may not be sold and possibly a concession will be given to an investor for a limited period (possibly 10 to 20 years)

The economic evaluation for Option 4 is detailed in Table 6.

Table 6. Economic evaluation for Option 4

	Status	From/incurred costs, LE	To/return, LE
Impact	Waste volume at site (tons)	200,000	99,000
	Occupied land (feddans)	3.7	3.7
	Maximum estimated gas emissions (megagrams/year)	3,500	1,700
	Estimated leachate generation (m^3/year)	17,000	8,500
	Incineration	Exists	Eliminated
	Risk of accidents	High	Eliminated
	Disamenity	Exists	Reduced
Evaluation of externalities	Cost of mortality	Exists	Reduced*
	Cost of morbidity	Exists	Reduced*
	Cost of lowered agricultural yield	Exists	Reduced*
	Cost of devegetation	Exists	Reduced*
	Cost of damage to buildings	Exists	Eliminated
	Cost due to climate change	Exists	Reduced*
	Cost of ecosystem degradation	Exists	Reduced*
Estimation of internal costs and return	Removal and transportation of 67,000 ton (50 LE x 67,000)	3,350,000	
	Capping and LMS (150 LE x 4200 x 3.7 + 300,000 LE)	2,630,000	
	Construction of transfer station	2,220,000	
	Remediation and monitoring**	500,000	
	Material recovery (20 LE x 33,000)	–	660,000
	Sale or rental value of land	–	5,000,000
Net	**−3,040,000 LE *** + reduced to eliminated externalities**		

*Depends on the level of remediation applied to site
**Cost of remediation is uncertain as data has to be collected to assess the level of degradation
***Although the project resulted in a negative net value, the land will return to the government after the concession period ends.

Table 7. Summary of the evaluation of decommissioning options

	Option	Total cost (LE)	Expected return (LE)	Net (LE)	External cost
1	Complete removal - no waste mining	18,220,000	20,000,000	1,780,000	Reduced to eliminated
2	Complete removal - waste mining	10,720,000	21,400,000	9,280,000	Reduced to eliminated
3	Partial removal – capping	12,350,000	5,000,000	−7,350,000	Reduced
4	Partial removal – capping – waste mining	8,700,000	5,660,000	−3,040,000	Reduced

6 Conclusions

Ill-defined features of the solid waste management (SWM) system in Egypt for a long period of time lead to aggravated health, environmental, economic, and social problems. The problem of giant uncontrolled open dumps of MSW is one the problems posing serious risk to the public health and the environment. Various solutions for decommissioning and removing such dumps are discussed in details and it was shown that best solutions, economically and environmentally, are based on practices that integrate the decommissioning project with an efficient solid waste management system.

The Shebin El-Kom open dump was studied and various closure options were compared. It has been shown that decommissioning the dump can be turned to a profit generating project.

The previous sections have shown expected cost or return from each of the evaluated options. The conclusions of this study help the decision maker to select the best decommissioning option and to clarify some facts beyond the numbers. Table 7 summarizes the results of the financial evaluation presented for decommissioning of the Shebin El-Kom dump. Significant increase on the return was achieved when waste is mined for compost. Material recovery or waste mining has a double economic effect, it reduces the amount of waste/reject to be transported to a landfill and it returns with a value for the recovered material. In addition, there is a combined environmental-economic benefit gained by reducing the footprint at the landfill. In this regard, options 2 and 4 are favorable decommissioning options.

It is also important to clarify that the capping options may result in a negative net amount (no return); nonetheless, the land will return to the government by the end of the concession period. There are high uncertainties regarding the concession premium value assumed in the economic evaluation. Therefore, the capping option may not be favored when compared with Option 2. Based on the above discussion and the summary of the evaluation, Option 2 reasonably stands as a favorable option that satisfies both the environmental and financial aspect of the project.

In order to proceed, the site requires more detailed site investigations and assessments including surveying of the site; and geotechnical, hydrological, and environmental investigations in order to fine tune the proposed budget and the expected return on the project. A precise feasibility study can follow. The same assessment methodology can be applied to other open dumping locations which have grown to similar scales. The key action is to integrate the closure or the decommissioning project with a wise solid waste management practice.

Acknowledgments. The Author would like to express their gratitude to Prof. Ahmed Gaber, Professor of Chemical Engineering, Faculty of Engineering, Cairo University and CEO of Chemonics Egypt Consultants, for supporting this study.

References

COWI: A Study on the Economic Valuation of Environmental Externalities from Landfill Disposal and Incineration of Waste. European Commission, DG Environment, October 2000

Chen, D., Zhai, X., Zhou, G.: Life cycle assessment of RDF production from aged MSW and its utilization system. In: Proceedings of the International Conference on Sustainable Solid Waste Management, Chennai, India, 5–7 September 2007, pp. 406–414 (2007)

Hogland, W., Hogland, M., Marques, M.: Enhanced landfill mining: material recovery, energy utilization and economics in the EU (Directive) perspective. In: Proceedings International Academic Symposium on Enhanced Landfill Mining, Houthalen-Helchteren, pp. 233–247 (2011)

Qian, X., Koerner, R.M., Gray, D.H.: Geotechnical Aspects of Landfill Design and Construction, p. 147. Prentice Hall (2002)

Zaki, T.: Country Report on the Solid Waste Management in Egypt. SWEEP-Net (2014). http://www.sweep-net.org/sites/default/files/EGYPT%20RA%20ANG%2014_1.pdf. Accessed May 2015

Zaki, T., Kafafi, A., Mina, M.B., Abd El-Halim, A.M., Saber, M.: New center for Integrated studies of Land & Environment (NILE). In: Annual Report for Solid Waste Management in Egypt, 2013, NSWMP/EEAA/GIZ (2013)

UNEP: Training Module: Closing of an Open Dumpsite and Shifting from Open Dumping to Controlled Dumping and to Sanitary Landfilling. United Nations Environment Programme (2005)

Geosynthetics, an Emerging Engineered Technology in Soil Stabilization in the Third World Countries for Sustainable Development

K. C. Onyelowe[1(✉)] and B. V. Duc[2]

[1] Department of Civil Engineering, Michael Okpara University of Agriculture, Umudike, P.M.B.7267, Umuahia 440109, Abia State, Nigeria
konyelowe@gmail.com, konyelowe@mouau.edu.ng

[2] Faculty of Civil Engineering, Hanoi University of Mining and Geology, Hanoi, Vietnam

Abstract. The technical review paper on Geosynthetics, an emerging engineered technology in foundation soil stabilization in the third world nations for sustainable development has tried to elucidate on the need to employ Geosynthetics in solving our challenging soil foundation and environmental geotechnics problems. Geosynthetics are engineered materials artificially produced for soil strength improvement. They appear in various forms applied to Environmental Geotechnics. There have been records of erosion manes and pavement failures in the south-south and southeastern regions of Nigeria and government agencies in these affected regions of the country collect security vote meant to be used to check this natural environmental problems, but lose sight of their responsibility for lack of focus and greed. It is strongly believed that this review paper will be an eye opener to both professionals and those in governance to know that there are cheaper and sustainable solutions to our engineering problems only if we are prepared to embrace what the technological trend of geoengineering has provided the third world nations. Furthermore, this work has elucidated the extent to which weak and expansive soils could be stabilized and possibly check gully erosion disasters destroying the regions by using geosynthetic materials and technology at the nanoscale.

Keywords: Geosynthetics · Emerging engineered technology
Soil stabilization · Sustainable development · Third world nation

1 Introduction

Nano-geosynthetics are nanoscale geosynthetic materials applied to the electroosmotic phase to improve weak engineering soil (Ahn et al. 2009). The third world nations are supposedly the center or atmosphere for development and human capacity building, but contrary to this opinion, they have refused to embrace the emerging trends in both engineering and technological development and human capacity building. Nigeria for example and in the wake of the oil boom of the 1950s and 1960s, it was generally perceived and believed that the government was to make use of the economy at the

H. Ameen and T. Sorour (Eds.): GeoMEast 2018, SUCI, pp. 38–43, 2019.
https://doi.org/10.1007/978-3-030-01929-7_3

time to boost technology by building factories and industries. Though to an extent, it did, viz; Ajaokuta Steel Complex, Kogi State, NAFCON Nigeria Ltd, PortHarcourt, Nigeria, Textile Mill Industries, Aba, Nkalagu Cement Industry, Eleme Petroleum Refinery, PortHarcourt, Petroleum Refinery, Kaduna and Warri, International Glass Industry, Aba, The Eastern States Golden Guinea Breweries, Umuahia, Modern Ceramics, Umuahia, Unilever Nigeria Ltd, Lagos and Aba, and many more. It is heartbreaking to note that of all mentioned industries, companies and factories, none is functional as of today. Let me rhetorically ask, what is wrong? Why are our seaports and airports, yet alive when they should have collapsed with the above mentioned unfortunate establishments? The third world nations have failed to manage earlier investments, how then do we talk about emerging trends? It is very obvious that we are our own problems and bane of the nation we all represent though the story is different in the Republic of Ghana, Chad, etc. today. Out of our greed and avarice, we have failed the people who look up to us for sustainability. Then I ask again, where are we headed? When other countries of the world are emerging and many emerged at all levels, what is our lot in the trend of development?

Human capacity building is a function of sustainable development, which is to say that if the third world nations have to build human capacity, they must have to, first of all, emerging trends for sustainable development. How do we go about it?

(a) Revitalize all failed industries and factories, because these projects didn't fail because we lacked the manpower or resources to maintain them, but because we allowed them to, maybe for the benefits we derive from importation thereby impoverishing the populace and jeopardizing the future of our unborn generation.
(b) Probe the trend of development and copy examples from the emerged nations and work towards making our nations sustainable by building industries and factories to produce products and materials that conform to the demands of the third world nations.
(c) To rain manpower at all levels to take up the responsibility of managing, running, working, serving, etc., at various capacities to project the dream of an emerging nation.

2 Foundation Soil Stabilization Emerging Technologies

There have been rampant cases of structural failures across the West African sub-region, mostly recorded in my beloved country, Nigeria. In the past five years, the cities of Umuahia, Owerri, Awka, Uyo, Abuja, Lagos, Ibadan and PortHarcourt have recorded gruesome cases of building collapse. Furthermore, our roads fail every other day few months after use and over 80% of these records resulted from foundation failures.

Equally, there is a popular structure called "7-Deck" located at #105 Faulks Road, Aba, Abia State, Nigeria, which was abandoned over 20 years ago after its completion to roof level because it showed vertical deflection of over 5°. This, according to a technical report by the Nigerian Society of Engineers, Aba branch, was as a result of differential settlement which went on to show that there was no proper site investigation prior to the construction which led to the colossal waste (Aba Engineer 2002).

The listed cases have resulted from negligence on the part of our professionals. In addition, like I have pointed out earlier that our highways fail few months after construction and use, who is to be blamed for all of these? The government takes a bigger proportion of the blame. In a situation where government agencies after fulfilling the conditions of the so called "due process" awards contract to an establishment and yet expects some percentage of this contract sum as what is popularly called "PR", the contractor in turn cuts corners in a bid to make profit ignoring trends that could give the project what it deserves.

There are emerging trends in soil stabilization, which, when tapped could place the third world nations at par with their counterparts in the second and first world nations. Over the years we have used metallic chlorides, lime (quick and hydrated) etc. (Onyelowe and Okafor 2012a; Onyelowe 2012), slurry, slag, additives and admixtures (bagasse ash, quarry dust, etc.) (Onyelowe and Okafor 2012b; Onyelowe et al. 2012; Onyelowe and Okafor 2013), but the trend based fundamentally on some environmental factors has changed.

In 2011, we published a work on the need to use geosynthetics to improve soil foundations for both roads (highways) and buildings and other civil engineering works (Onyelowe 2011). The third world nations would be better for it if nano-Geosyntheitcs and Geosynthetics are adopted to improve soils found to be insufficient in their load bearing capacity.

2.1 Geosynthetics

According to (Holtz 2001), Geosynthetics are planar products manufactured from polymeric materials (the synthetic) used with soil, rock or other Geotechnical-related materials as part of a civil engineering project or system. In 1970, there were only five or six Geosynthetics available, while today more than 600 different geosynthetic products are sold throughout the world (Holtz 2001; Berg and Suits 2010).

2.2 Geosynthetic Classifications

According to (Barthurst 2006), there are Geotextiles, geogrids, geonets, geomembranes, nano-geocomposites, geosynthetic clay liners (GCLs), geopipes, geocells, and geofoams used according to the problem to be solved and the prevailing environmental conditions. In fact, Geosynthetics are widely used in various areas of Geotechnical engineering amongst which includes stability and anchorage systems on slopes (Girard et al. 2005).

Oil was discovered and explored in Nigeria in the 1950s. However, the Geosynthetics are produced from polymeric materials which are products of petroleum. What would it have caused Nigeria to be the first country in the world to produce Geosynthetics and nano-Geosynthetics and use the same or probably export to other countries of the world if not negligence? The present generation would have built on the legacy of our past leaders and perhaps bequeath to the future.

a. ***Views from Advanced Economies***

According to IGS, these emerging technologies are updated from time to time to keep track of developmental flaws and strategies. In the USA and Brazil, for example, there are organizations established by government and individuals dedicated to the scientific and engineering development of society (Arulrajah et al. 2016).

b. ***Views of the Third World Nations; Nigeria***

In the third world nations, government and professionals complain of finances to execute projects. You can imagine a region where the use of Geosynthetics to save and improve soils proven to be poor in terms of strength to carry the anticipated load, for instance places like Lagos, PorHarcourt, Yenegoa, Onitsha, Asaba, Eket, etc. will always need soil improvement but often times negligence takes precedence to actuality. Secondly, Nigeria has all it takes to produce Geosynthetics here rather than importation of the same.

3 Foundation Soil Stabilization Emerging Technologies and Sustainable Development in Nigeria

Sustainable development is the development that meets the needs of the present without compromising the ability of the future generation to meet their own (Hoy et al. 2016) Sustainable development refers to a model of human development in which resource use aims to meet human needs while ensuring the sustainability of natural systems and the environment, so that these needs can be met not only in the present but also for generations to come (Phetchuay et al. 2014). When the third world nations are able to employ emerging trends in technologies, we can all record sustainable development, we can all bequeath to the future generation. Soil foundation failures, for example, have been the problem of the south-south, southeast and southwestern parts of Nigeria. Negligence in the proper procedures in foundation engineering and improvement of weak soils has led to this ugly trend. Development can only be sustained when proper and adequate procedures are followed considering emerging technological trends as it relates to environmental changes.

4 Foundation Soil Stabilization Emerging Technologies and Human Capacity Building in Nigeria

Human capacity building is the transformation a community approaches to social and environmental problems by assisting them being competent around skills (Phetchuay et al. 2016). According to (Phetchuay et al. 2014; Hoy et al. 2016), it is referred to as blended learning approach while (Suksiripattanapong et al. 2015; Sukmak et al. 2013), defines it as an unprecedented learning approach. Like we earlier to point out that "human capacity building is a function of sustainable development", when there is poise on the part of both government, corporate bodies, non-governmental organizations and individuals to develop sustainable environment, there will be cogent need to train manpower at all levels of developmental endeavor to take up such a great responsibility. Negligence to these procedures has left the environment in a state despair in terms of facility foundation failures. Soil improvement has advanced in the world into the realm of nano-geosynthetic and nanomaterials for use in concrete and admixture for soil stabilization and these have to be sustained to save our environment. In this light, experts in this field have to be trained and developed as well as encouraged to make the right decisions in terms state of the art procedures to solving our environmental problems.

5 Contribution from Government and Corporate Organizations

The government is yet to contribute towards sustainable development. Government that tries to build capacity without equal consideration to develop is wasting resources. What contribution have corporate bodies made? It is unfortunate that every facet of our nation has failed in this project. We must have to return to status quo, start all over and build a sustainable development that will meet the demands of human capacity building in an advanced world where geopolymers, geosynthetics, nanotechnology, nanomaterials and nano-geosynthetic materials could solve our ailing environmental problems with ecofriendly ease devoid of CO_2 emission and at a considerable cost.

6 Conclusion

The third world nations have a long way to go because the little distance we all think we have covered all these years has been a waste of time and resources. Why? Because, we have to start all over, learn from the mistakes of the past; borrow leaves from emerged nations of the world and do things right. To achieve this, we must drop ego, greed and avarice with our consciousness raised towards serving the cause of all humanity. Nano-geosynthetic materials have been developed locally to save our environment, e.g. nano palm bunch ash, nano kaolin, nano pozzolan, nano tyre ash, and lots more and have been used in the laboratory as fillers in the production of concrete and admixtures in soil stabilization. Results have shown improved soil admixture interface, improved soil strength, improved soil density and improved the soil California bearing ratio, which are the Geotechnical properties that certify an engineering soil suitable for use as subgrade and subbase material. Indulgence into this area and sustainable approach will not only save our ailing environment, but place Nigeria in the league of technologically advanced nations.

References

Aba Engineer: Technical paper on ethical practice for Engineers in Aba, Nigerian Society of Engineers, Aba Branch, vol. 21 (2002)

Ahn, K.J., Jeong, K., Lee, J.: Characteristics on electroosmosis ground improvement using nano-geosynthetics. J. Korean Geoenviron. Soc. **10**(4), 59–63 (2009)

Arulrajah, A., Kua, T.A., Horpibulsuk, S., Phetchuay, C., Suksiripattanapong, C., Du, Y.J.: Strength and microstructure evaluation of recycled glass-fly ash geopolymer as low-carbon masonry units. Constr. Build. Mater. **114**, 400–406 (2016)

Barthurst, R.J.: Geosynthetic classification. International Geosynthetic Society, USA (2006)

Berg, R.R., Suits, L.D.: Geosynthetics, A2K07: Committee on Geosynthetics, NY (2010)

Girard, H., Haddane, K., Poulain, D.: Stability and anchorage of geosynthetic systems on slopes: development of a new designing tool. In: International Workshop on Hydro-Physico-Mechanics of Landfills, France (2005)

Holtz, R.D.: Geosynthetics for soil reinforcement, the 9th Spencer J. Buchanam Lecture, Texas (2001)

Hoy, M., Horpibulsuk, S., Rachan, R., Chinkulkijniwat, A., Arulrajah, A.: Recycled asphalt pavement–fly ash geopolymers as a sustainable pavement base material: Strength and toxic leaching investigations. Sci. Total Environ. **573**, 19–26 (2016)

Onyelowe, K.C.: Geosynthetics and geotechnical properties of soil in a developing country: a lesson for Nigeria. Electron. J. Geotech. Eng. **16**, 1481–1487 (2011). ISSN 1089-3032. www.ejge.com

Onyelowe, K.C.: Soil stabilization techniques and procedures: a clue for the developing world-Nigeria. Glob. J. Eng. Technol. **5**(1), 65–69 (2012). www.enggtech.org

Onyelowe, K.C., Okafor, F.O.: A comparative review of soil modification methods. ARPN J. Earth Sci. **1**(2) (2012a). www.arpnjournals.com/jes

Onyelowe, K.C., Okafor, F.O.: Geochemistry of soil stabilization. ARPN J. Earth Sci. **1**(2) (2012b). www.arpnjournals.com/jes

Onyelowe, K.C., Okafor, F.O., Nwachukwu, D.G.: Geophysical use of quarry dust (as Admixture) as applied to soil stabilization and modification-a review. ARPN J. Earth Sci. **1**(1) (2012). www.arpnjournals.com/JES

Onyelowe, K.C., Okafor, F.O.: Portland cement/quarry dust improvement of olokoro laterite for road base. World J. Eng. Sci. **1**(4), 133–143 (2013). www.wjesonline.com

Phetchuay, C., Horpibulsuk, S., Arulrajah, A., Suksiripattanapong, C., Udomchai, A.: Strength development in soft marine clay stabilized by fly ash and calcium carbide residue based geopolymer. Appl. Clay Sci. **127**, 134–142 (2016)

Phetchuay, C., Horpibulsuk, S., Suksiripattanapong, C., Chinkulkijniwat, A., Arulrajah, A., Disfani, M.M.: Calcium carbide residue: alkaline activator for clay–fly ash geopolymer. Constr. Build. Mater. **69**, 285–294 (2014)

Suksiripattanapong, C., Srijumpa, T., Horpibulsuk, S., Sukmak, P., Arulrajah, A.: Compressive strengths of water treatment sludge-fly ash geopolymer at various compression energies. Lowl. Technol. Int. **17**(3), 147–156 (2015). www.europa.eu. Accessed 5 Apr 2016

Sukmak, P., Horpibulsuk, S., Shen, S.L., Chindaprasirt, P., Suksiripattanapong, C.: Factors influencing strength development in clay–fly ash geopolymer. Constr. Build. Mater. **47**, 1125–1136 (2013)

Geo-Environmental Impact of Bauxite Residue Stabilized with Fly Ash and Gypsum to Mitigate the Leaching Problems

D. Kishan(✉), S. S. Kushwaha, and N. Dindorkar

Department of Civil Engineering, MANIT, Bhopal 462003, MP, India
kishandhyd@gmail.com

Abstract. Red mud is one of the major wastes produced from aluminum industry which utilize bauxite (Al_2O_3) as raw material to extract aluminum out of it. This research work investigates the suitable utilization for a particular red mud samples depending upon its permeability and hazardous characteristics. Red mud was stabilized with Fly ash which is another solid waste material produced from thermal power plants and in combination with gypsum. The Fly ash content was varied from 0%, 15% and 25% and gypsum content is varied from 0%, 0.25% and 0.75% of dry weight of the red mud. The main objective of this research work is to evaluate the environmental impact of red mud utilization and explore the means to allay the leaching problem. The effect on the toxic metals in the leachate coming out from the compacted mix when fly ash is added solely or when combined with gypsum to red mud is to be clarified. All the experiments were carried out on each mix to obtain an optimum mix. The permeability and leachate characteristics are studied for different flowing periods from 1 to 7 days of each sample combination. An attempt further was made to understand the metal concentration in the leachate of raw red mud, red mud – fly ash and red mud-flyash–gypsum combinations. It is revealed from the test results and analysis that stabilization with proper proportioning the hazardous metal concentration and permeability of the red mud will decrease with increase in flow period.

Keywords: Leachate · Permeability · Red mud · Stabilization

1 Introduction

Depending on the source of bauxite used in aluminum industries, red mud which is the waste product of this process may contain different toxic metals and may contaminate the ground water. When water flows through red mud fill, it may carry toxic metals and may contaminate surface water sources also. The presence of toxic metals may not be considerable sometimes but biomagnifications of these metals may aggravate the situation. Thus, these toxic metals emanating from red mud may be responsible for different health hazards of surrounding habitats. There are different means available to mitigate leaching problems. Thus stabilization of red mud with fly ash, gypsum may be

H. Ameen and T. Sorour (Eds.): GeoMEast 2018, SUCI, pp. 44–56, 2019.
https://doi.org/10.1007/978-3-030-01929-7_4

one of the most useful method to mitigate the environmental problems, leaching and dusting. Through stabilization process, the exposed surface area can be reduced along with encapsulation of metals, to minimize the leaching of metals. In the present investigation, an attempt has been made to study the leachate properties of stabilized red mud by using fly ash and gypsum.

On the basis of economy consideration, fly ash was selected as stabilizing material small percentages of gypsum was also used to enhance strength of the matrix and for better encapsulation of heavy metals from Leachate. Subrat Kumar Rout et al. had determined the physical and chemical properties of red mud and made a comparative study of red mud with the soil and fly ash. A.R. Hind et al., studied the chemical properties of red mud. A chemical analysis would reveal that RM contains major constituents: silica, aluminum, iron, calcium, titanium. The minor constituents: Na, K, Cr, Ni, Ba, Cu, Mn, Pb, Zn etc. Y. Zhang, et al., studied the incoherence and high content of fine sand and coarse silt determine the importance of permeability research of red mud. As per the Author, it is weakly permeable with k $2.57 \times 10e{-5}$ and $3.62 \times 10e{-5}$ cm/s. The filtration coefficient of red mud affected by penetrating water tends to increase gradually because of leaching of easily soluble salts and flocculation of Ca++ lead to rapid decrease of salts. Semra Coruh et al., studied Use of fly ash, phospho gypsum and red mud as a liner material for the disposal of hazardous zinc leach residue waste. In the study, leachability of heavy metals from the zinc leach residue has been evaluated by mine water leaching procedure and toxicity characteristic leaching procedure (TCLP).

2 Methodology

2.1 Material Used

Red Mud - For the present work the red mud was collected from HINDALCO Ltd, Uttar Pradesh. The geotechnical properties of red mud like specific gravity (3.02), plasticity index (13.7), compaction characteristics (OMC = 35.0, MDD = 1.57 gm/cc), Unconfined compression strength (0.123 MPa), permeability (5.786e−7 cm/s). It can be seen that the red mud is highly alkaline with PH value of 11.3 and the specific gravity (3.02) is also very high compared to soil (usually 2.65). The high specific gravity of red mud is due to presence of iron rich minerals. As per Indian Standard Soil Classification System (IS 1498-1970) it can be classified as silt of medium compressibility (MI).

Fly Ash - For the present research work, low lime fly ashes (class F) are used and these were collected from Kota Thermal Power Plant, Kota, Rajasthan. The raw fly ash was collected in dry state from electro static precipitator and generally called as raw fly ash at the plant which has low pozzolanic reaction.

Gypsum - Analytical quality, anhydrous gypsum is being used. Gypsum is used to accelerate the pozzolanic reaction between Red mud and fly ash. The analytical quality was chosen to avoid the interference of impurities which may retard the initial hydration process.

2.2 Mix Proportions

The effect of variation of fly ash and gypsum on the red mud is studied for different aspects. Nine samples of varying fly ash and gypsum content by dry weight of red mud, as shown in Table 1 were prepared and tested for permeability and leachate characteristic.

Table 1. Sample combinations

S. No.	Sample combination
1	RM+0FA+0G
2	RM+0FA+0.25G
3	RM+0FA+0.75G
4	RM+15FA+0G
5	RM+15FA+0.25G
6	RM+15FA+0.75G
7	RM+25FA+0G
8	RM+25FA+0.25G
9	RM+25FA+0.75G

RM - Red mud, **FA**-Fly ash, **G**-Gypsum. (Numerals in the combination stand for the % by dry weight of red mud).

2.3 Permeability Characteristic

Depending on the sample proportions, various ingredients (red mud, fly ash and gypsum) were mixed thoroughly in dry state. Then required amount of water (OMC) was sprinkled over the dry sample and mixed systematically with hand. Then the sample was compacted in the permeability mould following the specification of IS: 2720 (part-7) - 1987 (Proctor light compaction method). After compaction the permeability moulds kept for variable head permeameter apparatus to find permeability and leachate analysis of the test specimens. After saturation of the test specimens the permeability was measured by falling head method, according to IS: 2720 (part-17) - 1987. Water was allowed to flow through the test specimens continuously for seven days and the readings were recorded daily.

The permeability test results for all the test specimens are presented in Figs. 1, 2 and 3, as the variation of permeability with flow periods for test specimens containing 0.0, 0.25 and 0.75% gypsum presented respectively. The graph 1–3 and Tables 2, 3 and 4 presented below are given for different combinations of test samples. The graphs are given for three gypsum contents for each sample composition. Figures 1, 2 and 3 illustrate the variation of permeability with flow periods for specimens compacted at optimum moister content (OMC) and flow period for 7 days respectively. Permeability test results are also presented tabular form in Tables 2, 3 and 4 for 7 days flowing period respectively.

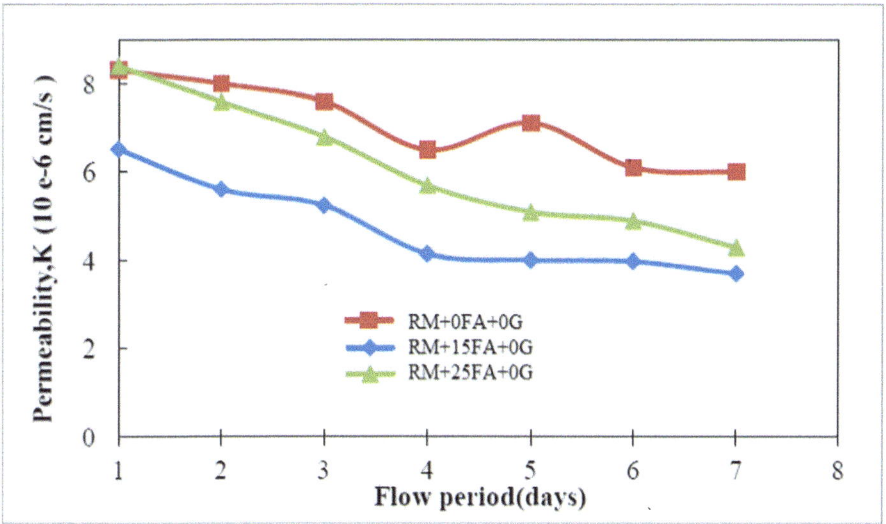

Fig. 1. Permeability vs flow period for stabilized red mud samples prepared with 0G

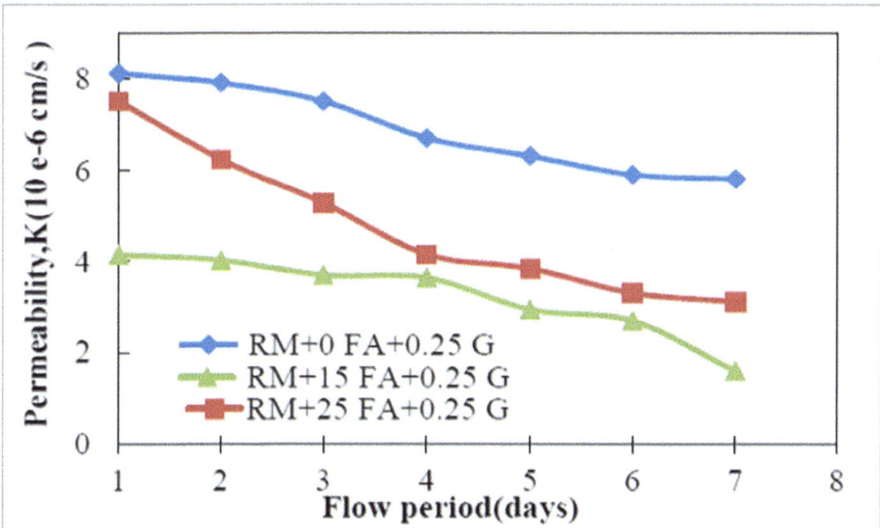

Fig. 2. Permeability vs flow period for stabilized red mud samples prepared with 0.25

Figures 1 present the effect of fly ash content on permeability for 7 days flow period samples containing 0.0% gypsum respectively. Figure 2 present the effect of fly ash content for 7 days flow period of the samples containing 0.25% gypsum respectively. Figures 3 present the effect of fly ash content on permeability for 7 days flow period samples containing 0.75% gypsum respectively. It is revealed that with increase in the

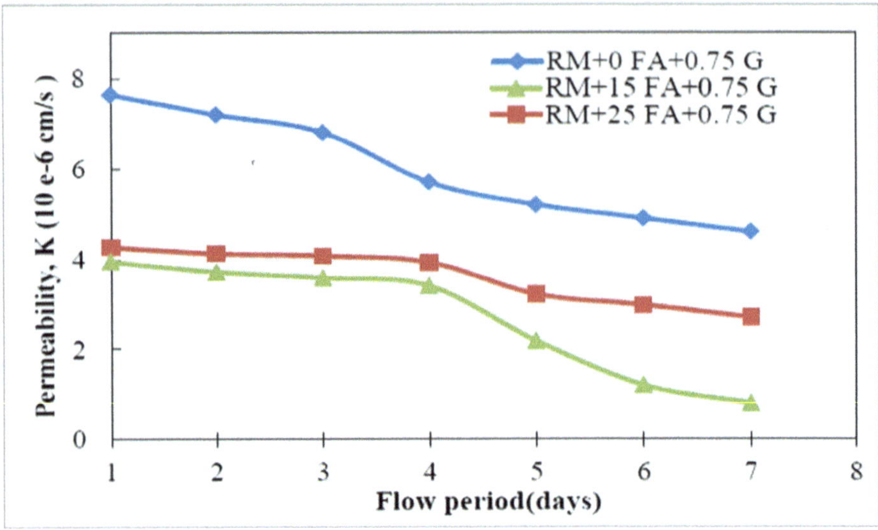

Fig. 3. Permeability vs flow period for stabilized red mud samples prepared with 0.75G

Table 2. Permeability, K (10e–6 cm/sec) of sample with FA & 0% G

Flow period (days)	Combination		
	A	B	C
1	8.32	6.52	8.41
2	8.01	5.61	7.60
3	7.60	5.24	6.80
4	6.50	4.15	5.70
5	7.12	4.01	5.10
6	6.10	3.98	4.90
7	6.00	3.70	4.30

Note:-
A = **RM+0FA+0G**
B = **RM+15FA+0G**
C = **RM+25FA+0G**

fly ash content there is reduction in permeability except for 1 day flow of 25% fly ash sample. But for the fly ash content 15% it was give the less permeability value. But from Figures with increase in flow periods, it is observed that the reduction in permeability to maximum extend in the present research. This may be due to the slow pozzolanic reaction; increase in flowing period helps the fly ash to participate in the pozzolanic reaction. Figures 1, 2 and 3 shows the effect of fly ash content on permeability for 7 days flowing period of all specimens. It is observed that the rate of decrease in permeability with fly ash depends on the presence of gypsum. This reduction of permeability may be due to reduction in void space and or pozzolanic reaction in presence of fly ash.

Table 3. Permeability (K × 10e–6 cm/sec) of samples with FA & 0.25% G

Flow period (days)	Combination		
	A	B	C
1	8.12	6.52	8.41
2	7.91	5.61	7.60
3	7.50	5.24	6.80
4	6.70	4.15	5.70
5	6.31	4.01	5.10
6	5.90	3.98	4.90
7	5.81	3.70	4.30

Note:-
A = **RM+0FA+0.25G**
B = **RM + 15FA+0.25G**
C = **RM+25FA+0.25G**

Table 4. Permeability (K × 10e–6 cm/sec) of samples with FA & 0.75% G

Flow period (days)	Combination		
	A	B	C
1	7.64	3.94	4.26
2	7.20	3.71	4.12
3	6.80	3.58	4.07
4	5.70	3.40	3.92
5	5.20	2.18	3.21
6	4.90	1.20	2.97
7	4.60	0.80	2.70

Note:-
A = RM+0FA+0.75G
B = RM+15FA+0.75G
C = RM+25FA+0.75G

Results presented for samples shows that addition of small percentages of gypsum (0.25% and 0.75%) along with fly ash has decreased the permeability compared to lime addition only. It is also observed that addition of 0.75% gypsum can reduce the permeability to a greater extent compared to the increase of gypsum content from 0.0% to 0.25%. This reduction of permeability due to the addition of gypsum may be due to the active participation of gypsum in red mud-fly ash interaction. Hence with the addition of small percentages of gypsum, optimum sample may be achieved from economy point of view as well as to minimize the permeability. It is revealed from Figs. 1, 2 and 3 for all sample combinations that with increase in flow period, there is reduction in permeability. Depending on sample proportion some of the test specimens show that the decreasing trend in permeability is followed by all the specimens. For all the sample combinations the effect of flow period on permeability is presented in Figs. 1, 2 and 3 for samples containing 0.0%, 0.25% and 0.75% gypsum respectively.

The trend is same for 7 days. This reduction in permeability with flow of water may be due to the formation of pozzolanic reaction of products presented in the mixes may reduce the inter connectivity of the voids. The fluctuation of permeability at the beginning of flow for 25% fly ash samples may be due to the lack of proper saturation of samples or washing of some soluble or loose particles and simultaneously formation of pozzolanic reaction products.

2.4 Leachate Analysis of Stabilized Red Mud

In this investigation water was allowed to passing through compacted specimens and the water coming out from the samples was analyzed for metals. Here the testing procedure very close to the field condition where water will passage through compacted red mud media and subsequently join the ground water. The effluent coming out from the outlet of the Permeability mould was collected in sampling bottle. The bottles were washed with acetone and distilled water before collecting the samples. To prevent volume change and evaporation, the samples were kept in refrigerator and analysis was done by using PC based Double Beam Spectro Photo Meter without any delay. The lowest of the allowable limits for drinking water of World Health Organization (WHO, 1984), Guidelines for Canadian Drinking Water Quality (GCDWQ, Department of Health and Welfare, 1979) and National primary Drinking water standards of EPA, USA (1980), the Indian drinking water standards and European drinking water standards are compared for present research work.

This section presents the test results of leachate of metals coming out from compacted stabilized red mud with fly ash and gypsum samples. These leachates were collected as effluent during Permeability test. All leachate test results, i.e. concentration of metals (ppm), are presented in tabular forms for 7 days of flowing period. The effects of different factors on the trend of the leachate test results are summarized a little later in tabular format. Tables 5, 6, 7, 8, 9, 10, 11, 12 and 13 present the concentrations of metals like Fe, Na, Al, Si, Ca, As, Mn, Cr, Cu, Mg, Ni, Pb and Zn analyzed by using pc based double beam spectro photo meter for 1 to 7 days of flow. For all the samples results are presented for the samples prepared at OMC only.

Table 5. Metal concentrations in leachate (ppm) for RM+0FA+0G sample

S. No.	Flow period	Permeability, K(10e−6) in cm/s	Metal concentration in ppm												
			Fe	Al	Si	Na	Ca	Cr	Mg	As	Mn	Cu	Ni	Zn	Pb
1	1 Day	8.32	0.42	ND	ND	0.20	ND	0.27	0.30	0.21	0.35	ND	ND	0.25	ND
2	2 Day	8.01	0.44	ND	0.19	ND	0.12	ND	0.30	ND	0.34	ND	0.29	0.24	0.19
3	3 Day	7.60	0.41	0.21	0.15	ND	0.08	0.12	ND	ND	0.32	ND	ND	ND	0.18
4	4 Day	6.50	0.39	0.19	0.13	ND	0.02	ND	ND	ND	0.32	ND	0.26	0.22	ND
5	5 Day	7.12	0.34	0.17	0.12	ND	0.03	ND	ND	ND	ND	0.31	0.25	ND	ND
6	6 Day	6.10	0.31	0.15	0.11	ND	0.03	ND	ND	ND	ND	0.34	0.26	ND	ND
7	7 Day	6.00	0.31	0.14	ND	ND	ND	ND	ND	ND	ND	ND	0.24	ND	ND

Table 6. Metal concentrations in leachate (ppm) for RM+15FA+0G sample

S. No.	Flow period	Permeability, K(10e−6) in cm/s	Metal concentration in ppm												
			Fe	Al	Si	Na	Ca	Cr	Mg	As	Mn	Cu	Ni	Zn	Pb
1	1 Day	6.52	0.40	ND	ND	ND	0.39	ND	ND	0.20	ND	0.15	0.28	ND	ND
2	2 Day	5.61	0.38	0.22	ND	ND	0.14	0.27	0.30	ND	0.34	0.14	ND	ND	0.18
3	3 Day	5.24	0.37	ND	0.20	ND	ND	0.20	ND	ND	0.31	0.12	0.23	0.19	0.21
4	4 Day	4.15	0.34	ND	0.19	ND	ND	ND	0.28	ND	0.32	0.11	0.21	0.20	ND
5	5 Day	4.01	0.35	ND	0.17	ND	0.02	ND	ND	ND	ND	ND	0.21	0.18	0.20
6	6 Day	3.98	0.31	ND	0.17	ND	0.03	ND	ND	ND	ND	ND	ND	0.16	ND
7	7 Day	3.70	0.31	ND	0.15	ND	0.03	ND	ND	ND	ND	ND	0.21	0.17	ND

Table 7. Metal concentrations in leachate (ppm) for RM+25FA+0G sample

S. No.	Flow period	Permeability, K(10e−6) in cm/s	Metal concentration in ppm												
			Fe	Al	Si	Na	Ca	Cr	Mg	As	Mn	Cu	Ni	Zn	Pb
1	1 Day	8.41	0.36	ND	ND	ND	0.21	0.45	ND	0.17	ND	ND	ND	ND	ND
2	2 Day	7.60	0.38	ND	0.18	ND	ND	ND	ND	ND	ND	ND	0.25	ND	0.23
3	3 Day	6.80	0.38	ND	0.17	ND	ND	ND	ND	ND	0.17	0.12	0.26	0.17	0.21
4	4 Day	5.70	0.37	ND	0.17	ND	ND	ND	ND	ND	ND	0.14	ND	0.16	0.19
5	5 Day	5.10	0.36	ND	0.14	ND	ND	0.24	ND	ND	ND	ND	0.25	0.15	0.22
6	6 Day	4.90	0.35	ND	0.12	ND	0.17	ND	ND	ND	ND	ND	ND	0.11	ND
7	7 Day	4.30	0.29	ND	0.12	ND	0.13	ND	ND	ND	ND	0.05	ND	ND	ND

Table 8. Metal concentrations in leachate (ppm) for RM+0FA+0.25G sample

S. No.	Flow period	Permeability, K(10e−6) in cm/s	Metal concentration in ppm												
			Fe	Al	Si	Na	Ca	Cr	Mg	As	Mn	Cu	Ni	Zn	Pb
1	1 Day	8.12	0.41	ND	ND	0.13	ND	ND	ND	ND	0.34	ND	0.24	ND	ND
2	2 Day	7.91	0.39	ND	0.18	0.12	ND	ND	ND	ND	0.33	ND	0.23	ND	0.35
3	3 Day	7.5	0.37	ND	0.15	ND	ND	ND	ND	ND	0.32	ND	ND	ND	0.14
4	4 Day	6.7	0.37	0.19	0.14	ND	0.07	ND	ND	ND	ND	0.31	0.25	ND	ND
5	5 Day	6.31	0.36	0.18	0.19	ND	0.06	0.20	ND	ND	ND	0.29	0.25	ND	ND
6	6 Day	5.9	0.34	0.16	0.16	ND	0.03	ND	ND	ND	ND	0.29	ND	ND	ND
7	7 Day	5.81	0.31	0.15	0.14	ND	0.02	0.20	ND	ND	ND	0.27	0.21	ND	ND

Table 9. Metal concentrations in leachate (ppm) for RM+15FA+0.25G sample

S. No.	Flow period	Permeability, K(10e−6) in cm/s	Metal concentration in ppm												
			Fe	Al	Si	Na	Ca	Cr	Mg	As	Mn	Cu	Ni	Zn	Pb
1	1 Day	4.15	0.36	ND	0.21	ND	ND	ND	ND	ND	ND	ND	0.25	ND	ND
2	2 Day	4.03	0.35	ND	0.22	ND	ND	0.26	0.32	ND	0.30	ND	0.24	ND	0.17
3	3 Day	3.70	0.33	ND	0.21	ND	ND	ND	ND	ND	0.28	ND	ND	ND	0.17
4	4 Day	3.64	0.34	ND	0.23	ND	ND	0.24	0.30	ND	0.28	0.12	ND	ND	0.12
5	5 Dav	2.95	0.32	ND	ND	ND	0.15	ND	ND	ND	ND	0.09	0.22	ND	0.09
6	6 Day	2.70	0.30	ND	0.21	ND	0.17	ND	ND	ND	ND	0.15	ND	ND	ND
7	7 Day	1.60	0.29	ND	0.24	ND	0.17	ND	0.31	ND	ND	ND	ND	ND	ND

Table 10. Metal concentrations in leachate (ppm) for RM+25FA+0.25G sample

S. No.	Flow period	Permeability, K(10e−6) in cm/s	Metal concentration in ppm												
			Fe	Al	Si	Na	Ca	Cr	Mg	As	Mn	Cu	Ni	Zn	Pb
1	1 Day	7.51	0.40	ND	0.18	ND	ND	ND	ND	ND	0.24	0.21	0.15	ND	ND
2	2 Day	6.24	0.38	ND	0.17	ND	ND	0.27	ND	ND	ND	0.23	0.13	ND	ND
3	3 Day	5.27	0.37	ND	ND	ND	ND	ND	ND	ND	0.23	ND	0.11	ND	0.15
4	4 Day	4.15	0.39	ND	ND	ND	ND	ND	ND	ND	ND	ND	0.13	ND	0.18
5	5 Day	3.84	0.35	ND	0.16	ND	0.17	ND	ND	ND	ND	0.16	0.12	ND	0.14
6	6 Day	3.31	0.32	ND	0.13	ND	0.13	0.26	ND	ND	ND	0.08	ND	ND	ND
7	7 Day	3.12	0.31	ND	0.09	ND	0.18	0.23	ND	ND	ND	ND	ND	ND	ND

Table 11. Metal concentrations in leachate (ppm) for RM+0FA+0.75G sample

S. No.	Flow period	Permeability, K(10e−6) in cm/s	Metal concentration in ppm												
			Fe	Al	Si	Na	Ca	Cr	Mg	As	Mn	Cu	Ni	Zn	Pb
1	1 Day	7.64	0.39	ND	ND	0.12	ND	ND	ND	ND	0.34	0.31	0.22	ND	ND
2	2 Day	7.2	0.38	ND	0.17	0.08	0.08	ND	ND	ND	0.32	ND	0.21	ND	ND
3	3 Day	6.8	0.38	ND	0.15	ND	0.07	ND	ND	ND	0.31	ND	0.23	ND	ND
4	4 Day	5.7	0.37	0.2	0.13	ND	0.05	ND	ND	ND	ND	0.29	ND	ND	ND
5	5 Day	5.2	0.35	0.2	0.12	ND	0.02	0.2	ND	ND	ND	0.28	ND	ND	ND
6	6 Day	4.9	0.32	0.2	0.1	ND	0.01	0.21	ND	ND	ND	0.24	0.21	ND	ND
7	7 Day	4.6	0.29	0.2	0.1	ND	0.05	0.18	ND	ND	ND	0.24	0.21	ND	ND

Table 12. Metal concentrations in leachate (ppm) for RM+15FA+0.75G sample

S. No.	Flow period	Permeability, K(10e−6) in cm/s	Metal concentration in ppm												
			Fe	Al	Si	Na	Ca	Cr	Mg	As	Mn	Cu	Ni	Zn	Pb
1	1 Day	3.94	0.29	ND	ND	ND	0.31	ND	ND	ND	ND	0.38	0.24	ND	ND
2	2 Day	3.71	0.27	ND	0.21	ND	0.33	ND	0.31	ND	ND	0.36	0.23	ND	0.18
3	3 Day	3.58	0.21	ND	0.22	ND	0.29	ND	ND	ND	ND	ND	0.24	ND	0.16
4	4 Day	3.40	0.18	ND	ND	ND	0.28	ND	0.34	ND	ND	0.33	0.21	ND	ND
5	5 Day	2.18	0.15	ND	0.23	ND	0.26	ND	ND	ND	ND	0.31	ND	ND	0.13
6	6 Day	1.20	0.12	ND	ND	ND	0.02	ND	ND	ND	ND	ND	ND	ND	0.11
7	7 Day	0.80	ND	ND	ND	ND	0.01	ND	ND	ND	0.27	0.26	ND	ND	ND
8	S Day	0.70	ND	ND	ND	ND	0.01	ND	ND	ND	0.27	0.26	ND	ND	ND

Table 13. Metal concentrations in leachate (ppm) for RM+25FA+0.75G sample

S. No.	Flow period	Permeability, K(10e−6) in cm/s	Metal concentration in ppm												
			Fe	Al	Si	Na	Ca	Cr	Mg	As	Mn	Cu	Ni	Zn	Pb
1	1 Day	4.26	0.41	ND	ND	ND	0.42	ND	ND	ND	ND	0.33	0.28	ND	ND
2	2 Day	4.12	0.40	ND	ND	ND	0.36	0.21	ND	ND	ND	0.29	ND	ND	0.21
3	3 Day	4.07	0.39	ND	0.17	ND	0.38	0.23	ND	ND	ND	ND	0.29	ND	0.19
4	4 Day	3.92	0.38	ND	0.15	ND	0.36	0.18	ND	ND	0.21	ND	ND	ND	ND
5	5 Day	3.21	0.40	ND	0.18	0.81	ND	0.15	ND	ND	ND	0.39	0.28	ND	0.15
6	6 Day	2.97	0.38	ND	0.14	ND	ND	ND	0.28	ND	0.22	ND	ND	ND	ND
7	7 Day	2.70	0.35	ND	0.12	ND	0.31	ND	ND	ND	ND	0.27	0.23	ND	ND

3 Discussion

3.1 Effect of Fly Ash Content

Figures 1, 2 and 3 present the effect of fly ash content on permeability for 7 days flow period samples containing 0.0% gypsum respectively. Figures 1, 2 and 3 present the effect of fly ash content for 7 days flow period of the samples containing 0.25% gypsum respectively. Figures 1, 2 and 3 present the effect of fly ash content on permeability for 7 days flow period samples containing 0.75% gypsum respectively. It is revealed that with increase in the fly ash content there is reduction in permeability except for 1 day flow of 25% fly ash sample. But for the fly ash content 15% it was give the less permeability value. The same trend is followed for all the samples with increase in flowing periods. From Figs. 1, 2 and 3 it is observed that for 7 days flowing periods with increase in fly ash content the permeability has decreased, maximum reduction of permeability occurs for addition of fly ash from 0.0% to 15.0%. But from Figures with increase in flow periods, it is observed that the reduction in permeability to maximum extend in the present research. This may be due to the slow pozzolanic reaction; increase in flowing period helps the fly ash to participate in the pozzolanic reaction. Figures 1, 2 and 3 shows the effect of fly ash content on permeability for 7 days flowing period of all specimens. It is observed that the rate of decrease in permeability with fly ash depends on the presence of gypsum. This reduction of permeability may be due to reduction in void space and or pozzolanic reaction in presence of fly ash.

3.2 Effect of Gypsum Content

Tables below presented for samples shows that addition of small percentages of gypsum (0.25% and 0.75%) along with fly ash has decreased the permeability compared to lime addition only. It is also observed that addition of 0.75% gypsum can reduce the permeability to a greater extent compared to the increase of gypsum content from 0.0% to 0.25%. This reduction of permeability due to the addition of gypsum may be due to the active participation of gypsum in red mud-fly ash interaction. Hence with the addition of small percentages of gypsum, optimum sample may be achieved from economy point of view as well as to minimize the permeability.

3.3 Effect of Flow Period

It is revealed from Figs. 1, 2 and 3 for all sample combinations that with increase in flow period, there is reduction in permeability. Depending on sample proportion some of the test specimens show that the decreasing trend in permeability is followed by all the specimens. For all the sample combinations the effect of flow period on permeability is presented in Figs. 1, 2 and 3 for samples containing 0.0, 0.25 and 0.75% gypsum respectively. The trend is same for 7 days. This reduction in permeability with flow of water may be due to the formation of pozzolanic reaction of products presented in the mixes may reduce the inter connectivity of the voids. The fluctuation of permeability at the beginning of flow for 25% fly ash samples may be due to the lack of proper saturation of samples or washing of some soluble or loose particles and simultaneously formation of pozzolanic reaction products.

4 Conclusions

The important conclusions drawn from this study are as follows

1. It is revealed that with increase in the fly ash content there is reduction in permeability except for 1 day flow of 25% fly ash sample. But for the fly ash content 15% it was give the less permeability value. But from Figures with increase in flow periods, it is observed that the reduction in permeability to maximum extend in the present research. This may be due to the slow pozzolanic reaction; increase in flowing period helps the fly ash to participate in the pozzolanic reaction.
2. It is also observed that addition of 0.75% gypsum can reduce the permeability to a greater extent compared to the increase of gypsum content from 0.0% to 0.25%. This reduction of permeability due to the addition of gypsum may be due to the active participation of gypsum in red mud-fly ash interaction. Hence with the addition of small percentages of gypsum, optimum sample may be achieved from economy point of view as well as to minimize the permeability.
3. It is observed from Tables 5, 6, 7, 8, 9, 10, 11, 12 and 13 that the concentration of Fe is increasing with increase in gypsum for all combinations of raw red mud. But with stabilization by fly ash the concentrations of Fe is decreasing and is in the slightly above the limits for drinking water standards based on drinking water guide lines of World Health Organization (WHO, 1984), the Indian drinking water standards and European drinking water standards. The Fe concentration is very low for 15FA and 0.75G combination as compared to the others.
4. The concentration of Al in leachate for raw red mud was increasing with flow period. After mixing with gypsum alone the Al concentration is slightly increasing
5. 0.14 ppm to 0.2 ppm with increase in flow period. But for stabilized red mud with fly ash and gypsum the Al was not determined in the leachate coming from the samples. The Al concentration in leachate was very nearer with drinking water standards irrespective of the sample proportion of raw red mud and flowing period.
6. The concentration of Si was decreased with increasing gypsum content and flowing period. But with stabilizing fly ash the Si concentration was increasing randomly. It may be due to the solubility of Si which is present in the fly ash.
7. The concentration of Na in leachate coming out from unstabilized red mud was slightly high, but with stabilization the concentration of Na in the leachate has not determined.
8. The concentration of Ca in leachate of raw red mud was very less. But by stabilization of red mud with fly ash the Ca concentration is increasing but the concentration value for Ca was below the allowable limit for drinking water quality but for some of the samples at a few flow periods, the concentration is altered randomly.
9. The Cr concentration in leachate for raw red mud was slightly decreased with increasing flow period and gypsum content. But with addition of fly ash only the Cr concentration in leachate was increased and the value was higher than the allowable limit for drinking water quality. But for some samples of 0.75G the Cr content is not determined.

10. The concentration of Mg in leachate coming out from unstabilized red mud was high, but with stabilization of red mud, the concentration of Mg in the leachate has decreased to very low value compared to drinking water standard irrespective of the sample proportion and flowing period. This may be due to the formation of Mg $(OH)_2$ in presence of low lime in fly ash.
11. The concentration of As in leachate was found at a 1 day flow periods of raw red mud and red mud stabilized with fly ash alone. This concentration exceeds the allowable limit for drinking water standards. But by addition of gypsum to stabilized red mud with fly ash the As was not found irrespective of flow periods.
12. The concentration of Mn in the unstabilized red mud is higher. By stabilizing the red mud and with increase in flow period the Mn content was decreased below the maximum allowable limit for drinking water quality.
13. The concentration of Cu was decreased with increasing gypsum content and flowing period. But with stabilizing fly ash the Cu concentration was decreasing upto 15% of fly ash and then increasing randomly for 25% of fly ash content.
14. The concentration of Ni in the unstabilized red mud for 7 days flow period samples is above the allowable limit compared with drinking water quality. But with increased in fly ash and gypsum content the Ni concentration was reduced slightly. But for some samples with different flow periods this was altered.
15. The concentration of Zn in leachate was found for raw red mud and red mud stabilized with fly ash alone. This concentration exceeds the allowable limit for drinking water standards. But by addition of gypsum to stabilized red mud with fly ash the Zn concentration was not found irrespective of flow periods.
16. The concentration of Pb in the unstabilized red mud for 7 days flowing period samples is above the allowable limit compared with drinking water quality. But With addition of gypsum, the concentration of Pb is reduced.

Acknowledgments. The tests were performed in the Geotechnical Engineering Laboratory of Civil Engineering Department, Mauluna Azad National institute of technology, Bhopal, India. The authors are thankful; to Civil Engineering department, Maulana Azad National institute of technology, Bhopal for providing laboratory facility.

References

Kalkan, E.: Utilization of red mud as a stabilization material for the preparation of clay liners. Eng. Geol. **87**, 220–229 (2006)

Agrawal, A., Sahu, K.K., Pandey, B.D.: Solid waste management in non-ferrous industries in India. Resour. Conserv. Recycl. **42**, 99–120 (2004)

Zhang, Y., Qu, Y., Wu, S.: Engineering geological properties and comprehensive of the solid waste (red mud) in aluminium industry. Environ. Geol. **41**, 249–256 (2001)

Hind, A.R., Bhargava, S.K., Grocott, S.C.: The surface chemistry of Bayer process solids: a review. Colloids Surf. A **V**(146), 359–374 (1999)

Renforth, P.: Contaminant mobility and carbon sequestration downstream of the Ajka (Hungary) red mud spill: the effects of gypsum dosing. Sci. Total Environ. **421–422**(2012), 253–259 (2012)

Bazdanis, G.: Removal of heavy metals from leachates using permeable reactive barriers filled with reactive organic/inorganic mixtures. In: Third International Conference on Environmental Management, Engineering, Planning and Economics, Skiathos island, Greece, 19–24 June 2011

Krishna, D.G.: Determination of Iron (II) and Iron (III) in presence of micellar medium using Ortho-hydroxy benzamide by spectrophotometry. Int. J. Inorg. Bioinorg. Chem. (2013). ISSN 2249-854

Suchita, R., et al.: An alternative to clay in building materials: red mud sintering using fly ash via Taguchi's methodology. Adv. Mater. Sci. Eng. **2013**, 7 (2013). Article ID 757923

Suchita, R., et al.: Neutralization and utilization of red mud for its better waste management. Arch. Environ. Sci. **6**, 13–33 (2012)

Wu, C., Liu, D.: Mineral phase and physical properties of red mud calcined at different temperature. J. Nanomater. **2012**, 6 (2012). Article ID 628592

Dodoo-Arhin, D., et al.: Fabrication and characterisation of Ghanaian bauxite red mud-clay composite bricks for construction applications. Am. J. Mater. Sci. **3**(5), 110–119 (2013)

Liu, D.-Y., Wu, C.-S.: Stockpiling and comprehensive utilization of red mud research progress. Materials **2012**(5), 1232–1246 (2012)

Singh, I.B., Chaturvedi, K., Morchchale, R.K., Yegneswaran, A.H.: Thermal treatment of toxic metals of industrial hazardous waste with fly ash and clay. J. Hazard. Matter. **141**, 215–222 (2007)

Bala, R.P., Arun, P., Chandra, S.A.: Study on CBR behavior of waste plastic (PET) on stabilized red mud and fly ash. IJSCR **2**(3), 232–240 (2013)

Hart, L.D.: Alumina Science and Technology Handbook Chemicals. The American Ceramic Society Inc., Westerville (1990)

Chaddha, M.J.: Asia pacific partnership of clean development and climate on industry perspectives-an overview. Jawaharlal Nehru Aluminum Research Development and Design Centre (2009)

Wang, P., Liu, D.-Y.: Physical and chemical properties of sintering red mud and Bayer red mud and the implications for beneficial utilization. Materials **2012**(5), 1800–1810 (2012)

Satyanayarana, P.V.V., et al.: Characterization of lime stabilized red mud mix for feasibility in road construction. Int. J. Eng. Res. Dev. **3**(7), 20–26 (2012)

Paramguru, R.K., Rath, P.C., Misra, V.N.: Trends in red mud utilization – a review. Miner. Process. Extr. Metall. Rev. **26**, 1–29 (2005)

Snars, K., Gilkes, R.J.: Evaluation of bauxite residues (red mud's) of different origins for environmental applications. Appl. Clay Sci. **46**, 13–20 (2009)

Subrat, K.R., Tapaswini, S., Sarat, K.D.: Utility of red mud as an embankment material. Int. J. Earth Sci. Eng. **5**(6), 1645–1651 (2012)

Evaluation of Surfactant Formulation for EOR in Some Depleted Oil Fields of Upper Assam

Miranda Kakoty and Subrata Borgohain Gogoi(✉)

Department of Petroleum Technology, Dibrugarh University, Dibrugarh, Assam, India
mirandakakoty@gmail.com, subrata@dibru.ac.in

Abstract. Enhanced Oil Recovery (EOR) techniques are becoming the thrust area to recover the residual oil in most of the matured depleted reservoirs. The microscopic displacement efficiency (E_D) of chemical Enhanced Oil Recovery (CEOR) particularly surfactant and/or alkali are dependent on the interfacial properties between the reservoir fluids, mainly crude oil and formation water. This paper discusses the details of all the results of the laboratory experiments carried out to develop an alkali-surfactant (A-S) EOR slug for two depleted oil fields of Barail formation of Upper Assam. A series of experiments such as IFT, adsorption study, compatibility study and thermal stability have been carried out to optimize an A-S formulation. Depending on the anionic nature of most reservoirs of Upper Assam, an anionic surfactant system comprising of a natural surfactant 'black liquor (BL)' and a synthetic surfactant 'Sodium Dodecyl Sulfate (SDS)' was adopted. This system at its critical micellar concentration (CMC) was mixed with alkalis: Na_2CO_3 and $NaHCO_3$ and corresponding CMC values at the lowest IFT were recorded. IFT experiments conducted separately with formation water (containing K^+, Ca^{2+}, Mg^{2+}) and with laboratory prepared brine (containing Na^+ only) established that IFT values are lower with formation water than that with the later. This study conducted in the temperature range from 30 °C to 80 °C showed that IFTs increase slightly with temperature, but within the range under study, temperature did not appear to have significant effect on IFT. Based on IFT reducing capabilities, a slug comprising of BL, SDS and Na_2CO_3 was selected for the EOR application in the considered depleted oil fields.

Thermal stability tests were carried out to evaluate the thermal decomposition tendency of the components. Both the surfactant system and alkali did not undergo any thermal degradation upto 90 days. The core flood experiments were carried out on core samples of the selected oil field with the optimized AS formulation following Buckley Leverette theory and calculations were done by Johnson, Bossler and Neumann (JBN) method. AS formulation thus developed in the laboratory was capable of lowering the oil water IFT to the range 10^{-2}–10^{-3} mN/m and the components were very much compatible with each other at reservoir temperature. Thermal stability study also produced expected results. The results of the core flood experiments were encouraging as 50–76% additional recovery has been achieved.

H. Ameen and T. Sorour (Eds.): GeoMEast 2018, SUCI, pp. 57–75, 2019.
https://doi.org/10.1007/978-3-030-01929-7_5

1 Introduction

Oil-production from EOR projects continues to supply an increasing percentage of the world's oil; about 3% of the worldwide production now comes from EOR (Taber et al. 1997). Various studies and reviews reveal that only a total of about 20–40% of Original Oil in Place (OOIP) can be recovered by employing primary and secondary recovery leaving behind more than 60% of OOIP unrecovered (Gogoi and Kakoty 2017; Demirbas et al. 2015; Alagorni et al. 2015; Muggeridge et al. 2014; Sandrea 2007). Chemical EOR methods including polymer, surfactant, micellar, emulsion and alkaline flooding hold a particular attraction for recovering the "residual oil" left in the reservoir after water flooding (which can produce oil until the residual waterflood oil saturation S_{or} is reached (Islam and Ali 1990; Das et al. 2017). The tertiary oil recovery is mainly dependent on the properties of oil/aqueous/formation interfaces viz., capillary forces, contact angle, wettability, viscous forces and interfacial tension. These properties are represented by a dimensionless group called the capillary number, N_c which is a measure of the occluded oil to enhance the oil recovery. The residual saturations decrease as N_c increases (Lake 1989; Alotaibi and Nasr-El-Din 2009). For better EOR efficiencies, the N_c has to be maximized by either increasing viscosity or by reducing IFT. Viscosity can be increased by flooding with chemical solutions of high apparent viscosity (polymers) while IFT is reduced by injection of surfactant solution or water with optimum salinity. Polymer allows a gain of about one order of magnitude in the N_c value and also provide a good mobility control but cannot alone produce the required change in IFT or capillary number. The decrease in IFT between crude oil and formation water lowers the capillary forces, facilitates oil mobilization, and enhances oil recovery (Das and Gogoi 2015a; Isehunwa and Olubukola 2012). Therefore, IFT of reservoir fluid system is an important thermo-physical property in investigating the capillary forces acting on trapped oil within the reservoir rocks.

Crude oil contains surface-active molecules such as low molecular weight fatty acids, naphthenic acids and asphaltene. Depending on pH and salinity, these natural surfactants will interact with the added synthetic surfactants in order to enhance the oil recovery (Trabelsi et al. 2012; Langevin et al. 2004). Alkali reacts with naturally occurring organic acid (naphthenic acid) in the crude oil and forms foam, which interact synergistically with added surfactant to produce ultra-low IFT. Ultra-low IFT is obtained through the distribution of surfactant between oil and water phase; and arrangement of surfactant at interface of oil/water. In addition, the alkali injected with surfactant can reduce surfactant adsorption and aid in lowering IFT (Phukan et al. 2018; Gogoi 2011; Al-Sahhaf et al. 2002; Gao et al.1995; Martin et al. 1985). The addition of polymer increases the viscosity of its aqueous phase so that the mobility of aqueous phase decreases, results in the decrease in mobility ratio (MR = Mobility of Displacing fluid/Mobility of Displaced Fluid) and increase in sweep efficiency. However, designing these aqueous chemical mixtures is complex, and must be tailored to the reservoir rock and fluid (i.e. crude oil and formation brine) properties of the application. A strategic design methodology can help to provide a low cost, well performing chemical formulation.

This paper describes a systematic laboratory approach to low cost, high performance chemical flooding and explores novel approaches to A-S flooding in reservoirs based on BL. The design strategy first uses IFT experiments to quickly select and optimize concentrations of injected chemicals. Formulations performing well in this experiment were subjected to aqueous stability, thermal stability and adsorption experiment respectively. Finally the selected formulation is validated in core flood experiment that adheres to necessary design criteria such as pressure and salinity gradients, surfactant adsorption and capillary effects.

2 Experimental Description

2.1 Apparatus

The surface tension (ST) of WHTC oils and water were measured by 'Easy Dyne Tensiometer K20' (Du Nouy Ring Method); the IFTs were measured by 'M6500 Spinning Drop Tensiometer'. Thermal stability test was carried out using a PerkinElmer-STA 8000 model thermogravimetric analyser (TGA). Adsorption experiment was carried out with the help of Shimadzu UV-1800 model UV-Visible Spectrophotometer. Porosity of Core sample was measured with Coretest Systems, Inc. TPI-219 model Teaching Helium Porosimeter and air permeability was measured with OFITE Model 360 Air Permeameter. Core flooding experiment was carried out using Gon Engineerung core flooding instrument. Formation water analyses were carried out in SYSTRONICS Water Analyzer 371 and SYSTRONICS Flame Photometer 128.

2.2 Materials

2.2.1 Chemicals and Fluids

Surfactants- Sodium Dodecyl Sulphate (SDS) and Black Liquor (BL). SDS procured from Merck Specialities Pvt. Ltd., Mumbai, with purity more than 90% (M.W. 288.37 g/mol) is a synthetic anionic surfactant with the formula $CH_3(CH_2)_{11}SO_4Na$. As a result of its hydrocarbon tail, and its anionic "head group", it has amphiphilic properties that allow it to form micelles, and so act as a surface-active compound. BL obtained as an effluent of Nogaon paper mill, Assam, India is a natural anionic hydrophilic surfactant. Its component Sodium lignosulfonate possesses very good surface-active properties to lower IFT between oleic and aqueous phase. Detailed specification and properties of BL is discussed (Gogoi and Das 2012).

Alkali- Sodium carbonate (Na_2CO_3) and Sodium bicarbonate ($NaHCO_3$). Na_2CO_3 Anhydrous Pure procured from Loba Chemie Pvt. Ltd. Mumbai, India with purity 99.5% (M.W. 105.99 g/mol) and $NaHCO_3$ from Avantor Performance Materials India Ltd. with purity 99% (M.W. 84.01 g/mol).

Crude oil, Formation water and synthetic brine- Four samples of wellhead tank crude (WHTC) oils M1, M2, N1 and N2; M1 and M2 from Moran Barail Upper Pool Block and N1 and N2 from Naharkatiya Main Barail sand Block of Upper Assam and corresponding formation waters were collected from Oil India Limited, Duliajan, Assam. Synthetic Soft brine (SFB) that mimics saline formation brine was prepared by

Sodium Chloride (NaCl) salt procured from Merck Specialities Pvt. Ltd., Mumbai, with purity 99% (M.W. 58.44 g/mol).

2.2.2 Core Samples

Core sample obtained from a depth of more than 3000 m from Moran (M) and Naharkatiya (N) oil fields of Upper Assam were used.

2.3 Procedure

The WHTC oils on which IFTs were measured was first separated from formation water. The separation includes decanting the oil, then vigorously mixing the oil with 2–4 drops de-oiler, then heating this mixture to 50 °C and finally allowing the mixture to settle.

2.3.1 IFT Determination

Aqueous solutions of SDS at 0.1–1.2% (w/w) in 2000 ppm brine and BL at 1–10% (w/w) in 2000 ppm brine were prepared; their IFTs with the WHTC oils were measured at 80 °C and corresponding CMC values were recorded. Then, in the BL surfactant solution (prepared by adding BL at its CMC value to 2000 ppm brine) SDS at concentrations 0.1–1.2% (w/w) were added and the IFTs of this surfactant mixture was measured at 80 °C. It was observed that even though this surfactant mixture lowered the IFTs to a low value of 0.12–0.239 mN/m, the required ultra-low IFTs for EOR was not obtained. Therefore, further experiments were conducted to examine behavior of the surfactant system as Alkali-Surfactant (A-S) slug for the two oil fields under study by adding alkali at its optimum value. To select the best alkali for A-S slug, two species of alkalis: Na_2CO_3 and $NaHCO_3$ were taken under consideration. Both these alkalis at different concentrations (1 wt% and 2 wt%) were added to the surfactant system and measurement of their respective IFTs was done.

All IFTs so far in this study were measured in brine solutions prepared in distilled water for meaningful comparison of surfactant slug meant for A-S EOR. In absence of any suitable surfactant that can impart ultra-low IFT to the reservoir crude-oil system, an attempt was made to at least measure the base IFT of the WHTC oils with their corresponding formation water and laboratory prepared brine. This compares the IFT behavior between the brine and actual formation water. The analysis of the formation waters were also carried out to validate the effect of the various ions present in the formation water for the reduction of IFT. This was accompanied by measuring IFT at the various temperatures upto 80 °C to examine the effect of temperatures on IFT.

The IFT between crude oil and water interface was measured with Spinning Drop Tensiometer at 30, 60 and 80 °C. The equilibrium IFT was measured from the equation

$$\gamma = 1.44 * 10^{-7} * d^3 * \theta^2 * \Delta\rho \tag{1}$$

Where,

γ IFT [mN/m]

d Diameter of the drop (LCD read directly from instrument) [mm]

θ Rotation (read directly from instrument) [rpm]
$\Delta\rho$ Density difference between crude oil and tested solution [g/cc]

2.3.2 Aqueous Stability Test

Aqueous stability assessment of the selected AS formulation from IFT experiment was carried out at salinities ranges from 2000 ppm to 10000 ppm. This test determines the clarity and homogeneity of all the dispensed aqueous solutions; also provides an upper value of salinity below which all chemical constituents can coexists in the aqueous phase. The AS formulation was dispensed in an array of glass vials having SFB from 2000 ppm to 10000 ppm, agitated and then allowed to settle at temperature 80 °C. Aqueous stability was assessed by visually inspecting any cloudiness and/or phase separation.

2.3.3 Thermal Stability Test

The thermal stability of the samples prepared for aqueous stability test was carried out by thermal gravimetric analysis. Simultaneous Thermal Analyzer (STA) 8000 determines the mass loss of surfactant with temperature. This test was conducted for a temperature range of 25 °C to 500 °C with a heating rate of 20 °C/min in a nitrogen environment.

2.3.4 Adsorption Test

Surfactant adsorption on reservoir rocks surface is one of the major factors that significantly reduce the effectiveness of a chemical flooding for enhanced oil recovery. The surfactant depletion method was used to determine the amount of adsorbed surfactant, and the analyses were conducted with the UV-Vis spectrophotometer.

Core sample M and N were crushed and then dried at 100 °C for 24 h. The dried crushed samples then screened through a 60-mesh ASTM sieve (0.25 mm aperture). One gram of undersized M and N sample of screening was mixed separately with 15 ml surfactant solution (with and without alkali) of different concentrations. The mixtures were shaken vigorously in magnetic shaker and then centrifuged at 1800 rpm for 10 min.

The supernatant liquid was filtered and analyzed. The amount of surfactant adsorbed was determined by measuring different concentrations in the solution before and after contact with crushed rock. The amount of adsorbed surfactant (A in mg surfactant/g rock) was calculated from:

$$A = V(C_o - C_e)/m \tag{2}$$

Where,

C_o Concentration before adsorption, mg/mL
V Solution volume, mL
C_e Equilibrium concentration (after adsorption), mg/mL
m Mass of adsorbent/particles, g

Details of procedure of adsorption on porous media by UV-Vis spectrophotometer is elaborated in (Johansen 2014).

2.3.5 Core Flooding

Sandstone core N was initially drilled and cut to about 3.5 inch (8.8 cm) length and 1.5 inch (3.8 cm) diameter; then dried. Before commencing fluid injection, porosity and air permeability of the core was determined. Then the core was wrapped with Teflon or PTFE tape and weighed; inserted into the core holder. The displacing fluids were inserted into the appropriate chamber of the core flooding instrument. Once, all the flow tubing connections were made and the system was checked for leaks, fluid injection was initiated.

The core was initially saturated with brine and brine was flooded through the core to measure the absolute permeability (k_w) of the core by steady state Darcy's law. It was then flooded with oil to measure the effective permeability of oil (k_o) till initial water saturation (S_{wi}) by the unsteady state JBN method following the Buckley Leverett assumptions (Buckley and Leverett 1942). The core is then flooded with secondary brine to measure the effective permeability of brine (k_w) till residual oil saturation (S_{or}) again by the unsteady state JBN method following the Buckley Leverette assumptions. The relative permeabilities (k_{rw} and k_{ro}) were calculated by dividing the effective permeabilities with absolute permeability of the core. Details of JBN method is discussed in (Mamudu 2016; Mamudu et al. 2017; Shikhov et al. 2017; Carrillo et al. 2008). The selected slugs were then flooded to recover the S_{or} and the calculations were done as per the unsteady state.

3 Results and Discussion

3.1 Chemical Formula Design Based on IFT

Chemical flooding methods are based on:

- Improving the mobility ratio i.e., making the mobility of the displacing fluid lesser than or equal to the mobility of the displaced fluid.
- Incrementing N_c through the reduction of interfacial tension (IFT) between the displacing and the displaced phases (as evident from Eq. 2) to values as small as possible.

The fundamental requirement of Chemical EOR processes is to produce low IFT. Surface tension (ST) and IFT of water and WHTC oils were measured and the resultant ranges are presented in Table 1.

Table 1. ST and IFT of water and WHTC oils at 40 °C

Sl. no	Description	ST or IFT, mN/m
1	Brine	68
2	Oil	10–40
3	Oil/Brine	5–20

During a waterflood, N_c given by equation no (2) is around and has to be increased by an order of at least two to three magnitudes to enable additional oil recovery (Iglauer et al. 2010; Abrams 1975; Green and Willhite 1998). Surfactant can increase N_c to or by lowering IFT to low or ultralow levels (i.e. 0.01–0.001 mN/m).

$$N_c = \frac{\mu V}{\gamma \cos \theta} \tag{3}$$

Where,

N_c Capillary number [–]
V Darcy velocity [m/s]
M Viscosity of injected phase [Pa.s]
Θ Contact angle [–]
Υ Interfacial tension oleic phase–aqueous phase [N/m]

The principal limitation of most chemical flooding is adsorption of surfactants on reservoir rock surface. As the clay of porous media of Upper Assam Basin is negatively charged (Hazarika et al. 2018; Das and Gogoi 2015b), anionic surfactant is preferred. The porous media repels the anionic surfactant and thus show low adsorption effects. This led to the selection of SDS and BL as surfactants for the porous media of Upper Assam Basin. BL is a highly toxic compound and can cause severe health hazards to humans and poses threat to aquatic life if disposed of without treatment. Thus, prior to discharge, it must undergo proper treatment, which in turn has high financial implications. Therefore, proper strategic use of BL for Surfactant EOR may not only provide economic recovery of residual oil but also save the environment from its hazards.

Figures 1 and 2 describe the IFT behavior of BL and SDS for N1, N2, M1 and M2 WHTC oil at 80 °C respectively. The reservoir temperature of Barail sand block is reported as 70–90 °C. CMC is the concentration at which the surfactant solution begins to form micelles in large amount and all additional surfactants added to the system go to micelles. CMC values of these surfactants at the lowest IFT for the entire WHTC oil sample were found out and are given in Table 2 along with their corresponding IFTs.

The investigation of IFT behavior of the surfactant system prepared by adding SDS at concentration range 0.1–1.2% (w/w) to BL solution (BL + 2000 ppm brine) was done. Figure 3 shows the IFT variation of surfactant system and CMC at the lowest IFT for each for WHTC oil.

It is seen from Table 2 that BL can lower the IFT of oil/brine to 1.2–1.8 mN/m from value of 5–20 mN/m (Table 1), while SDS can lower IFT to 0.834–1.2 mN/m for all the WHTC oils. The Fig. 3 reveals that instead of using BL and SDS alone, if the surfactant mixture is used, there is a further reduction of IFT to 0.12–0.239 mN/m. Although the surfactant system is found to be more effective in lowering IFT for all the tested WHTC oils as compared to that of BL and SDS alone; yet the required ultra-low IFT for an efficient EOR was not achieved. Xie et al. 2014, Thomas et al. 2016, Samanta et al. 2012 widely address the effectiveness of alkali injected in conjunction with surfactant in recovering the residual oil. Besides helping surfactant to produce ultra-low IFT, alkaline solution reacts with a divalent to form precipitates.

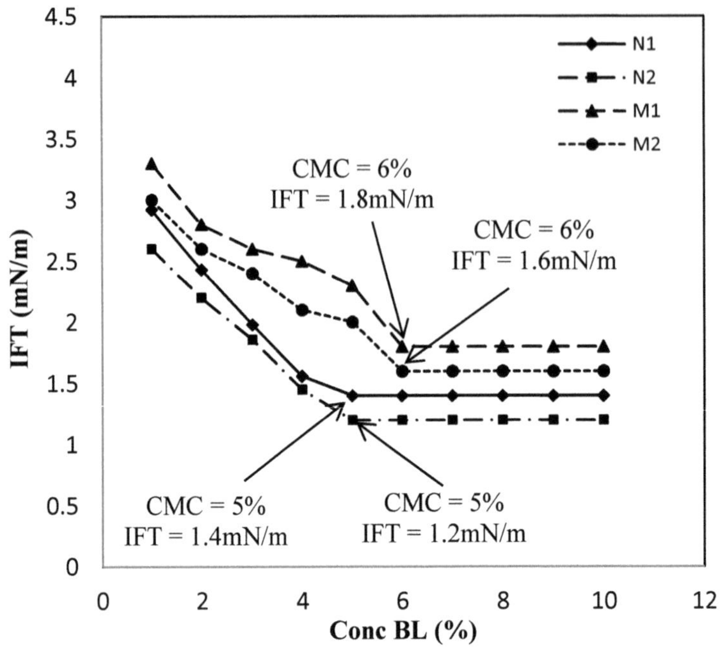

Fig. 1. CMC of BL at 80 °C for N1, N2, M1 and M2 WHTC oil

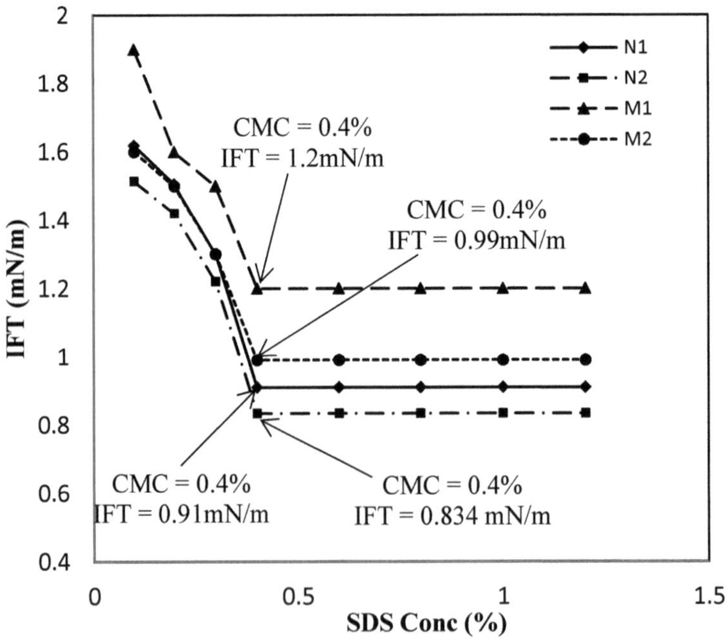

Fig. 2. CMC of SDS at 80 °C for N1, N2, M1 and M2 WHTC oil

Table 2. IFTs at CMC value of SDS and BL for WHTC oils at 80 °C

Sl no	Oil	Surfactant	CMC (w/w)	IFT (mN/m)
1	N1	BL	5	1.4
2	N2	BL	5	1.2
3	M1	BL	6	1.8
4	M2	BL	6	1.6
5	N1	SDS	0.4	0.91
6	N2	SDS	0.4	0.834
7	M1	SDS	0.4	1.2
8	M2	SDS	0.4	0.99

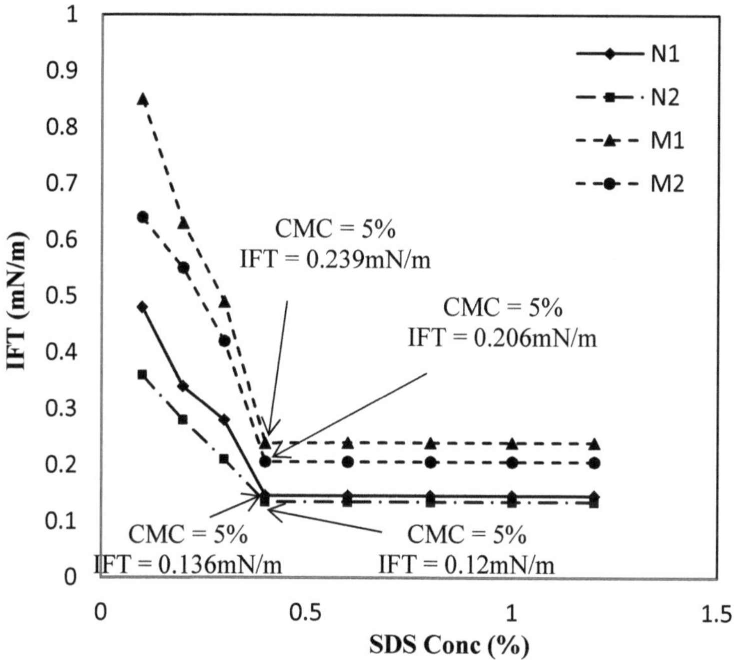

Fig. 3. CMC of surfactant system at 80 °C for WHTC oils

The precipitates preferentially reduce high permeability channels, thus sweep efficiency is improved. This process is known as mobility controlled caustic flood [MCCF] (Sheng 2015; Sarem 1974). Also, addition of alkalis increases pH, thus providing an optimal salinity for the surfactant slug and lowers the surfactant adsorption or retention on reservoir rock (Sheng 2015), optimizing the economic use of surfactant solution.

Therefore, further experiments to examine the effect of alkali in interfacial behavior of reservoir fluids at reservoir temperature were conducted. Na_2CO_3 and $NaHCO_3$ were added at different concentration (1 wt% and 2 wt%) to the formulated surfactant system

Table 3. Effect of alkalis on IFT

Sl. no.	Oil	Na_2CO_3 Conc. (wt%)	$NaHCO_3$ Conc. (wt%)	IFT (mN/m)
1	N1	0	0	0.146
		1	0	0.049
		2	0	0.012
		0	1	0.112
		0	2	0.102
2	N2	0	0	0.135
		1	0	0.052
		2	0	0.009
		0	1	0.105
		0	2	0.09
3	M1	0	0	0.239
		1	0	0.084
		2	0	0.035
		0	1	0.117
		0	2	0.103
4	M2	0	0	0.206
		1	0	0.062
		2	0	0.022
		0	1	0.128
		0	2	0.085

to study the change in IFT. Table 3 shows the IFT for the surfactant system and A-S slug recorded during this process.

It is evident from Table 3 that for N1 WHTC oil, addition of both the alkalis to the surfactant system results in lowering of IFT to lesser values as compared to the surfactant mixture without alkali. Adding Na_2CO_3 at 1 wt% concentration to the surfactant system brings down the IFT value from 0.146 mN/m to 0.049 mN/m. Increasing the Na_2CO_3 concentration to 2 wt% introduces further IFT reduction to a value of 0.012 mN/m. $NaHCO_3$ also reduces IFT as compared to the surfactant system alone, but the reduction is not much pronounced as found in the case of Na_2CO_3. Similar trend in reduction of IFT was seen in N2, M1 and M2 WHTC oils as well. It may be noted that in case of N2, the A-S slug having 1 wt% Na_2CO_3 showed an excellent reduction in IFT to a value of 0.009 mN/m at 80 °C. Thus, it may be summarized that using Na_2CO_3 (2 wt%) with aqueous surfactant system cause IFT reduction by a factor 10^{-1}–10^{-2} as compared to Surfactant mixture alone.

The oil/water IFT depends on multiple parameters such as reservoir temperature, pressure, crude composition, formation water salinity and composition to mention a few. All IFTs to this point were measured in brine solutions prepared in distilled water. The base IFT of the WHTC oils with their corresponding formation water and brine having almost the same salinity as that of formation water were carried out at 30, 60 and 90 °C. This compares the IFT behavior with respect to the laboratory prepared brine and actual formation water and establishes the effect of temperature on IFT. The outcome of this investigation is presented in Table 4.

Table 4. Base IFTs at various temperatures in formation water and brine

Sl. no	Well no	Temp °C	IFT in formation water (mN/m)	Formation water salinity as NaCl in (ppm)	IFT in 2000 ppm NaCl in DW (mN/m)
1	N1	30	4.34	1850	10.21
		60	5.57		13.12
		80	5.92		14.47
2	N2	30	3.07	1900	7.49
		60	4.23		10.01
		80	4.69		9.03
3	M1	30	7.18	2280	12.96
		60	7.78		13.42
		80	9.31		18.45
4	M2	30	8.24	2200	13.39
		60	8.39		14.32
		80	8.95		15.96

It is evident from Table 4, that crude IFTs in formation water increases with temperature. However, within the range of 30–80 °C, temperatures do not appear to have significant effect on IFT. Same is the case with the laboratory prepared brine. Crude IFTs are higher in brine than that in formation water. This may be due to the fact that in the laboratory prepared brine many of the other ions present in the formation water are absent. Table 5 shows the details found while characterizing the formation water under consideration.

Another aspect of this analysis is that the concentration of the divalent cations such as Ca^{2+} and Mg^{2+} will let us know about the scaling that can cause severe formation plugging. In all the formation water under study, concentration of Ca^{2+} and Mg^{2+} is <20 mg/l (ppm); hence, scale formation should not be a matter of concern (Donaldson et al. 1985).

3.2 Aqueous Stability

Aqueous stability is important to ensure a stable surfactant slug for injection, one where alkali and surfactant are compatible with electrolyte concentration at selected salinity. An experiment containing any phase separation or precipitation at aqueous phase beyond the selected salinity fails this screening.

Observations of the aqueous solutions were recorded first at room temperature and next the ampules were equilibrated at 80 °C using oven. The aqueous solutions were

Table 5. Formation water analysis data

Sl no	Characteristics	Unit	Formation water			
			N1	N2	M1	M2
1	Appearance	–	Turbid	Turbid	Turbid	Turbid
2	pH	–	8.4	8.5	8.55	8.32
3	Salinity as NaCl	mg/l	1850	1900	2280	2200
4	Total dissolved solid	mg/l	3140	3536	13118	11338
5	Turbidity	NTU	179	803	27	175
6	Chloride	mg/l	516	1152	1383	1334
7	Na^+	mg/l	2667	2000	3067	3333
8	K^+	mg/l	40	55	650	300
9	Ca^{2+}	mg/l	16	16	16	16
10	Mg^{2+}	mg/l	5	3	14.4	14.4

continued to monitor after 80 °C also. It was observed that aqueous solutions with the formulated A-S yield homogeneous and aqueous stable mixture up to 10000 ppm salinity. The aqueous stability has not changed after 60 days, demonstrating the stability of the AS system under the condition of high temperature and salinity.

3.3 Thermal Gravimetric Analysis

The thermal analysis of formulated surfactant is illustrated in Fig. 4. Three mass loss steps were observed. The first region, with a mass loss of 22% from ambient to 100 °C corresponds to the loss of weakly bonded water molecules (Carlino and Hudson 1998; Prinetto et al. 2000). Then, a 74.7% mass loss occurred sharply in the second region from 100 °C to 150 °C, revealing that A-S formulation's molecules start to decompose at temperatures exceeding 100 °C. The third degradation region from 150 °C to 500 °C represents a complex thermal decomposition with mass loss of 0.7%; the residual components of A-S formulation was thermally stable up to 500 °C (Carlino and Hudson 1995; Xi et al. 2005). But since the reported reservoir temperature of Barail sand block is 70–90 °C, A-S formulation would retain an average of 92.5% of its original structure and mass (Ibrahim et al. 2006). It can be concluded that the A-S formulation is thermally stable under the desired reservoir temperature for the application in EOR.

3.4 Adsorption

In the depletion method, the results being presented as isotherms which are simply plots of the amount of surfactant adsorbed per gram of solid or per surface area of solid versus the equilibrium surfactant concentration at a constant temperature. This is a measure of the extent of surface area of the adsorbent that is covered by the adsorbent molecules at a given condition. The models commonly used to describe anionic surfactant adsorption isotherm are monolayer Langmuir and empirical Freundlich model.

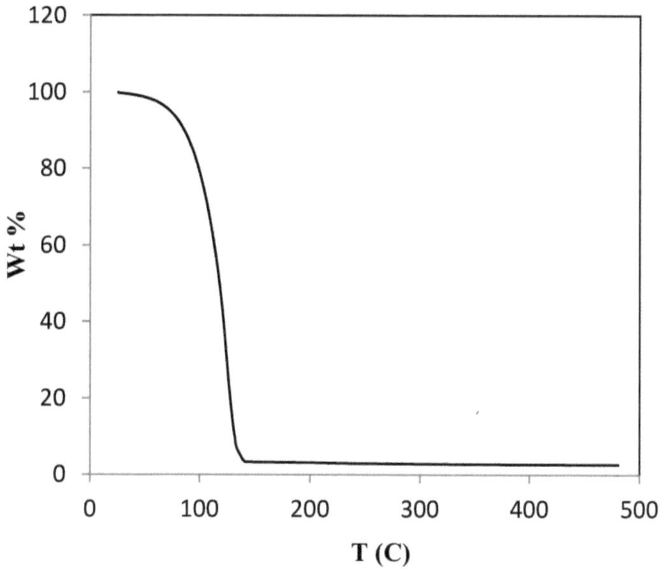

Fig. 4. Thermal stability curve for A-S formulation

The Langmuir equation is presented in Eq. (4) and its linearized form in Eq. (5)

$$\Gamma = \Gamma_{max} \frac{K_L C_e}{1 + K_L C_e} \tag{4}$$

$$\frac{1}{\Gamma} = \frac{1}{\Gamma_{max}} + \frac{1}{K_L \Gamma_{max}} \tag{5}$$

Where,

Γ Amount adsorbed mg/g
Γ_{max} Maximum adsorption capacity for forming monolayer mg/g
K_L Langmuir equilibrium constant L/mg
C_e Equilibrium aqueous concentration mg/L

The Freundlich equation is presented in Eq. (6) and its linearized form in Eq. (7)

$$\Gamma = k_f C_e^{\frac{1}{n}} \tag{6}$$

$$\log \Gamma = \log k_f + \frac{1}{n} \log C_e \tag{7}$$

Where,

Γ Amount adsorbed mg/g
k_f Constant for a given system (indicator of adsorption capacity)

n Constant for a given system (measure of the intensity of adsorption) L/g
C_e Equilibrium aqueous concentration mg/L

The experimental data are plotted on graph as in Figs. 5, 6, 7 and 8 for both the rock samples with surfactant mixture with alkali and without alkali; and investigated the suitable isotherm.

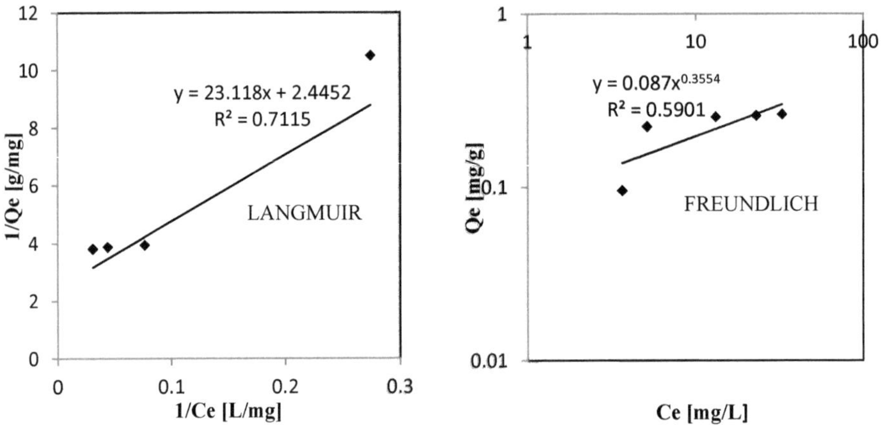

Fig. 5. Adsorption of S mixture (without alkali) on rock sample M

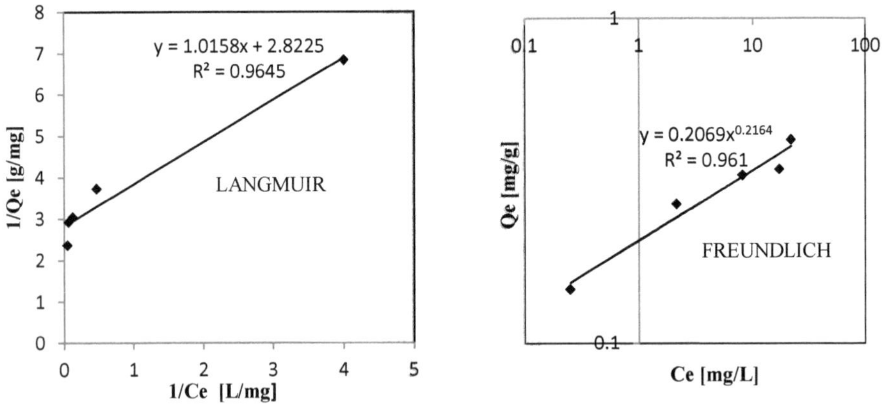

Fig. 6. Adsorption of S mixture (with Na_2CO_3) on rock sample M

Figures 5 and 6, display the Langmuir and Freundlich isotherms for the adsorption of Surfactant mixture and of formulated A-S on rock samples M respectively. In both the cases, experimental values match the Langmuir model better than Freundlich model. It is observed that the maximal adsorption in the isotherm has decreased when the alkali was introduced to the surfactant system. [Γ_{max} (Without alkali) = 0.409 mg/g; Γ_{max} (With alkali) = 0.347 mg/g]

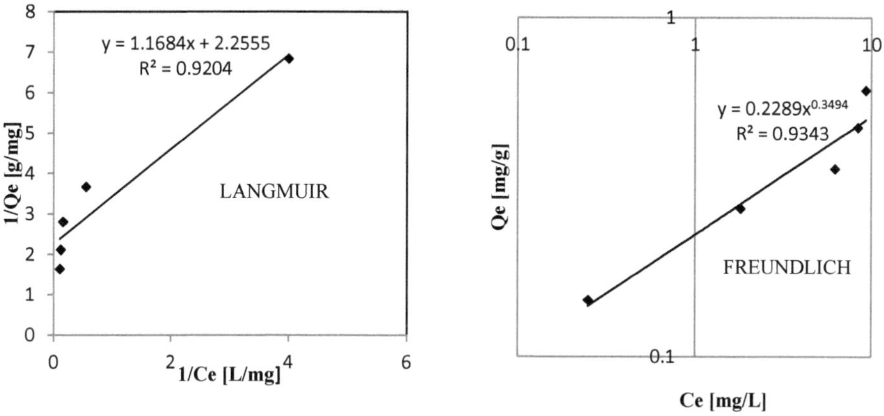

Fig. 7. Adsorption of S mixture (without alkali) on rock sample N

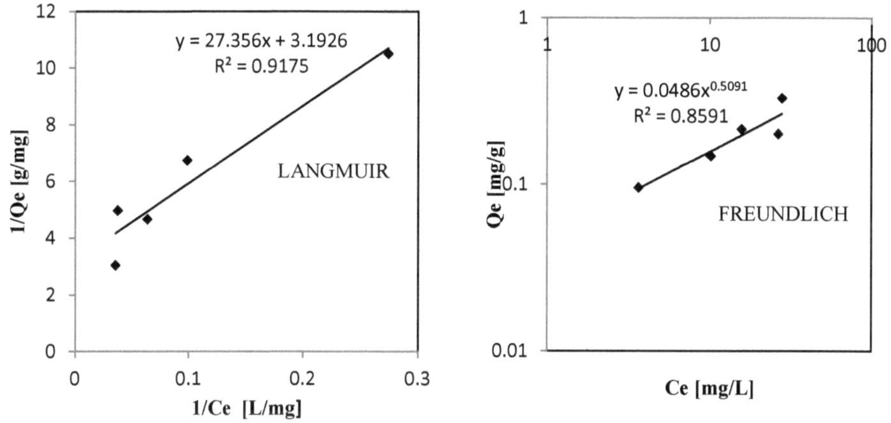

Fig. 8. Adsorption of S mixture (with Na_2CO_3) on rock sample N

Examination of plots in Fig. 7 suggest that both models has good fits for the adsorption of S mixture on rock sample N. Based on the R^2 values, the Freundlich model give a somewhat better match than the Langmuir model. Freundlich constants k_f and n are calculated to be 0.2289 L/g and 2.86, respectively. However, it is observed from Fig. 8 that experimental values for the adsorption of this S mixture containing alkali matched the Langmuir models better than Freundlich model with Γ_{max} = 0.313 mg/g.

3.5 Core Flooding

Core flood experiment is a means to validate whether a promising chemical formulation that performed well in the previous experiments is capable of recovering residual oil in

Table 6. Measured properties of core M during core flood

Mass, dimension	Mass (dry core only)	m (g)	224.64
	Total length	L (cm)	8.74
	Diameter	D (cm)	3.68
	Pore volume	PV (cc)	15.53
	Porosity	φ	16.69
Permeabilty	Air (dry, bare core)	k_{air} (md)	10.8
	Absolute (SFB)	k_{br} (md)	8.2
	Relative oil (at S_{wi})	k_{ro}^{o}	0.84
	Relative water (at S_{or})	k_{rw}	0.124
Saturation	Initial water	S_{wi}	0.124
	Initial oil	S_{oi}	0.876
	Residual oil	S_{or}	0.631

cores with low surfactant retention. The core flood experiment showed significant residual oil recovery of 75.5%. Table 6 lists dimension, permeability, and saturation core property data measured or calculated during core preparation and pre-flood (SFB flood, oil flood, water flood) experiments.

4 Conclusion

This work concludes that, based on IFT criteria, a composition of 5% BL (w/w), 0.4% SDS (w/w) and 2wt% Na_2CO_3 for depleted reservoirs of Naharkatiya; and 6% BL (w/w), 0.4% SDS (w/w) and 2wt% Na_2CO_3 for depleted reservoirs of Moran best suites as A-S slug for chemical EOR. This A-S slug lowers IFT to levels 10^{-3}–10^{-2}, which will lead to increase of N_c favorable for EOR. The presence of ions other than the Na^+ in the formation water enhance further reduction of IFT between the oleic and aqueous phases of the reservoir. IFTs of crude oil-formation water system increases slightly with temperature. However, the study indicates that temperature is not a pertinent parameter for lowering the IFT in case of the fields under study. As the Ca^{2+} and Mg^{2+} concentration in the formation water under consideration is less than 20 ppm, no scaling problem is expected for A-S EOR. The formulated A-S slug showed aqueous and thermal stability at reservoir temperature. Static adsorption experiment also revealed low adsorption of the formulation for both the core samples.

This novel formulation was tested in a core flood for Naharkatiya oil field and 76% of the water flood residual oil saturation was recovered. Procurement of core sample of desired dimensions from Moran oil field is under process and core flooding experiment on these samples will be conducted in our future work. This is a very significant development with a high potential to decrease the cost of chemical enhanced oil recovery since the primary surfactant used here is black liquor which is cheap and easily available as effluent from paper mill.

Acknowledgement. We gratefully acknowledge the financial assistance provided by University Grants Commission, New Delhi, India, under 21st Century Knowledge Initiative Program for Indo-US Project F. No. 194-1/2009(IC) dated 20.2.15 titled "Foam-assisted CO_2 Flooding for the depleted Reservoirs of Upper Assam Basin and in Candidate Reservoirs in Louisiana."

References

Abrams, A.: The influence of fluid viscosity, interfacial tension, and flow velocity on residual oil saturation left by waterflood. Soc. Pet. Eng. J., Soc. Pet. Eng. (1975). https://doi.org/10.2118/5050-PA

Alagorni, A.H., et al.: An overview of oil production stages: enhanced oil recovery techniques and nitrogen injection. Int. J. Environ. Sci. Dev. (2015). https://doi.org/10.7763/IJESD.2015.V6.682

Alotaibi, M.B., Nasr-El-Din, H.A.: Salinity of injection water and its impact on oil recovery. Paper SPE 121569, presented at EUROPEC/EAGE Conference and Exhibition, Amsterdam, The Netherlands, 8–11 June 2009. https://doi.org/10.2118/121569-ms

Al-Sahhaf, T., et al.: Producing ultralow interfacial tension at the oil/water interface. Pet. Sci. Technol. J. (2002). https://doi.org/10.1081/LFT-120003712

Buckley, S.E., Leverett, M.C.: Mechanism of fluid displacement in sands. Trans. AIME (1942). https://doi.org/10.2118/942107-G

Carlino, S., Hudson, M.J.: Thermal intercalation of layered double hydroxides capric acid into an Mg–Al LDH. J. Mater. Chem., R. Soc. Chemistry. (1995). https://doi.org/10.1039/jm9950501433

Carlino, S., Hudson, M.J.: A thermal decomposition study on the intercalation of Tris (oxalato) ferrate, trihydrate into a layered (Mg/Al) double hydroxide. Solid State Ion. (1998). https://doi.org/10.1016/S0167-2738(97)00500-6

Carrillo, E.R.P., et al.: A new method for the experimental determination of three-phase relative permeabilities. Ciencia, Tecnol. Y Futur. J., Colombian Petroleum Institute (ICP) - ECOPETROL S.A. (2008). ISSN 0122-5383

Das, B.M, Gogoi, S.B.: Relating IFT with oil recovery with special reference to bhogpara porous media of Upper Assam Basin. J. Pet. Eng. Technol. (2015a). ISSN 2231-1785

Das, B.M., Gogoi, S.B.: Adsorption study of a natural product into the oil producing porous media of Upper Assam Basin. Int. J. Res. Eng. Appl. Sci. (2015b). ISSN 2249-3905

Das, B.M., et al.: Micellar-polymer for enhanced oil recovery for Upper Assam Basin. Resour.-Effic. Technol. (2017). https://doi.org/10.1016/j.reffit.2017.01.003

Demirbas, A., et al.: Utilization of surfactant flooding processes for enhanced oil recovery (EOR). Pet. Sci. Technol. J. (2015). https://doi.org/10.1080/10916466.2015.1060503

Donaldson, E.C., et al.: Enhanced Oil Recovery. I. Fundamentals and Analyses. Elsevier, Amsterdam, The Netherlands (1985)

Gao, S., et al.: Laboratory investigation of combination of alkali/surfactant/polymer technology for Daqing EOR. SPE Reserv. Eng. (1995). https://doi.org/10.2118/27631-PA

Gogoi, S.B., Kakoty, M.: A study of CO_2 flooding on wave velocities in the Naharkatiya oil reservoir of Upper Assam Basin. Resour.-Effic. Technol. (2017). https://doi.org/10.1016/j.reffit.2016.12.011

Gogoi, S.B., Das, B.M.: Use of an effluent for enhanced oil recovery. Indian J. Chem. Technol. NISCAIR (2012)

Gogoi, S.B.: Adsorption – desorption of surfactant for enhanced oil recovery. Transp. Porous Media (2011). https://doi.org/10.1007/s11242-011-9805-y

Green, D.W., Willhite, G.P.: Enhanced Oil Recovery. SPE Textbook Series, vol. 6. SPE Publications (1998)
Hazarika, K., et al.: Characterization of crude oil for enhanced oil recovery: a study with anionic surfactant. Int. J. Ambient Energy (2018). https://doi.org/10.1080/01430750.2017.1421573
Ibrahim, Z.B., et al.: Laboratory aspect of chemical EOR processes evaluation for Malaysian oilfields. Paper SPE 100943, presented at Asia Pacific Oil and Gas Conference and Exhibition, Adelaide, Australia, 11–13 September 2006. https://doi.org/10.2118/100943-ms
Iglauer, S., et al.: New surfactant classes for enhanced oil recovery and their tertiary oil recovery potential. J. Pet. Sci. Eng. (2010). https://doi.org/10.1016/j.petrol.2009.12.009
Isehunwa, S., Olubukola, O.: Interfacial tension of crude oil-brine systems in the Niger Delta. Int. J. Res. Rev. Appl. Sci. **10**, 460–465 (2012)
Islam, M.R., Ali, S.M.F.: New scaling criteria for chemical flooding experiments. J. Can. Pet. Technol. (1990). https://doi.org/10.2118/90-01-02
Johansen, T.: Investigation of adsorption of surfactants onto kaolinite and relations to enhanced oil recovery methods. MS thesis, Norwegian University of Science and Technology, Trondheim, Norway (2014)
Lake, L.W.: Enhanced Oil Recovery. Prentice-Hall, California (1989)
Langevin, D., et al.: Crude oil emulsion properties and their application to heavy oil transportation. Oil Gas Sci. Technol. Rev. IFP Energ. Nouv. (2004). https://doi.org/10.2516/ogst:2004036
Mamudu, A.: An extension of Johnson, Bossler and Neumann JBN method for calculating relative permeabilities. Paper SPE 184492, presented at SPE Annual Technical Conference and Exhibition, Dubai, UAE, 26–28 September 2016. https://doi.org/10.2118/184492-stu
Mamudu, A., et al: Comparative approach to relative permeability predictions. Paper SPE 189174 presented at SPE Nigeria Annual International Conference and Exhibition, Lagos, Nigeria, 31 July–2 August 2017. https://doi.org/10.2118/189174-ms
Martin, F.D., et al.: Enhanced recovery of a 'J' sand crude oil with a combination of surfactant and alkaline chemicals. Paper SPE 14293, presented at SPE Annual Technical Conference and Exhibition, Las Vegas, Nevada, 22–26 September 1985. https://doi.org/10.2118/14293-ms
Muggeridge, A., et al.: Recovery rates, enhanced oil recovery and technological limits. Philos. Trans. A Math. Phys. Eng. Sci. J. (2014). https://doi.org/10.1098/rsta.2012.0320
Phukan, R., et al.: Enhanced oil recovery by alkaline-surfactant-alternated-gas/CO_2 flooding. J. Pet. Explor. Prod. Technol. (2018). https://doi.org/10.1007/s13202-018-0465-0
Prinetto, F., et al.: Synthesis and characterization of sol-gel Mg/Al and Ni/Al layered double hydroxides and comparison with co-precipitated samples. Microporous Mesoporous Mater (2000). https://doi.org/10.1016/S1387-1811(00)00197-9
Samanta, A., et al.: Comparative studies on enhanced oil recovery by alkali–surfactant and polymer flooding. J. Pet. Explor. Prod. Technol. (2012). https://doi.org/10.1007/s13202-012-0021-2
Sandrea, I.: Global oil reserves – recovery factors leave vast target for EOR technologies. Oil Gas J. **105**, 44–47 (2007)
Sarem, A.M.S.: Secondary and tertiary recovery of oil by MCCF process. Paper SPE 4901, presented at SPE California Regional Meeting, San Francisco, California, 4–5 April 1974. https://doi.org/10.2118/4901
Sheng, J.J.: Investigation of alkaline-crude oil reaction. Petroleum (2015a). https://doi.org/10.1016/j.petlm.2015.04.004
Sheng, J.J.: Status of surfactant EOR technology. Petroleum (2015b). https://doi.org/10.1016/j.petlm.2015.07.003

Shikhov, I., et al.: An experimental and numerical study of relative permeability estimates using spatially resolved T1-z NMR. Transp. Porous Media (2017). https://doi.org/10.1007/s11242-017-0855-7

Taber, J.J., et al.: EOR screening criteria revisited-part 1: introduction to screening criteria and enhanced recovery field projects. SPE Reserv. Eng. (1997). https://doi.org/10.2118/35385-PA

Thomas, N.C., et al.: Alkali and hybrid-alkali flooding as a tertiary oil recovery mode: prospects and challenges. Int. J. Pet. Petrochem. Eng. (2016). https://doi.org/10.20431/2454-7980.0202005

Trabelsi, S., et al.: Effect of added surfactants on the dynamic interfacial tension behaviour of alkaline/diluted heavy crude oil system. Oil Gas Sci. Technol. Rev. IFP Energ. Nouv. (2012). https://doi.org/10.2516/ogst/2012033

Xi, Y., et al.: Thermogravimetric analysis of organoclays intercalated with the surfactant octadecyl trimethylammonium bromide. J. Therm. Anal. Calorim. (2005). https://doi.org/10.1007/s10973-005-0750-2

Xie, D., et al.: Organic alkali for heavy oil chemical EOR improves the performance over inorganic alkali. Paper SPE 172895, presented at SPE International Heavy Oil Conference and Exhibition, Mangaf, Kuwait, 8–10 December 2014. https://doi.org/10.2118/172895-ms

Proportioning of CLSM Containing Class C Fly Ash, GGBS, Quarry Dust and Processed Slag Sand

C. P. Ramesh[1], H. P. Vageesh[2], T. Raghavendra[2(✉)], B. C. Udayashankar[2], and A. Shashishankar[3]

[1] Department of Civil Engineering, SET- JAIN University, Bengaluru, Karnataka, India
cp.ramesh@jainuniversity.ac.in
[2] Department of Civil Engineering, R. V. College of Engineering, VTU, Bengaluru, Karnataka, India
{vageeshhp, raghavendrat, udayashankar}@rvce.edu.in
[3] Department of Civil Engineering, AMC Engineering College, VTU, Bengaluru, Karnataka, India

Abstract. Controlled low strength materials (CLSM) is a backfill material which is neither concrete nor soil cement but it has properties similar to both. CLSM is a fluid mixture usually made of Portland cement, water, and fine aggregate or fly ash or both. However, other industrial by-products & recycled materials are also incorporated & encouraged as long as they are considered to be easily available, cost-effective for specific application along with the necessary characteristics of the mixture such as flow ability, strength, and density. Consistency of CLSM is similar to that of a slurry or lean grout, however several hours after placement the material hardens enough to bear loads without settling. Nowadays proper disposal of waste materials which are produced from various industries is a severe problem in many countries causing environmental hazards and raises the potential to contaminate water, air and soil resources and carbon emission. The present experimental research aims to arrive at suitable mix proportion combinations for CLSM by the incorporation of waste industrial by products such as Class C fly ash and GGBS (Ground-granulated blast-furnace slag) as partial replacement for cement. Quarry dust (QD) and Processed Slag Sand (PSS), was utilized in total replacement to natural river sand as fine aggregates, for proportioning of CLSM. Variations were considered in binder ratios and water contents. The fresh and hardened properties of CLSM in terms of flow, unconfined compressive strength were investigated. An attempt has been made to develop predictive flow and strength models known as phenomenological models to assist Engineers in the design of CLSM mixes with these material combinations at a future date or some other place. The findings from this study would benefit Engineers to adopt sustainable practice through incorporation of the above waste or by-products in CLSM – in large quantities unlike concrete or any other material.

Keywords: CLSM · PSS sand · GGBS · Class C fly ash

H. Ameen and T. Sorour (Eds.): GeoMEast 2018, SUCI, pp. 76–86, 2019.
https://doi.org/10.1007/978-3-030-01929-7_6

1 Introduction

Soil-cement has been a widely used material in geotechnical-engineering practices for a long time. CLSM is relatively new and is different from conventional soil cement. Compaction is usually required for soil-cement, which is not the case for CLSM. Fly ash and GGBS are wastes which are produced in huge quantity by industrial manufacturers such as power plants and steel manufacturing plants. Although these solid wastes have been used as a construction material for many applications, nearly 50–70% of fly ash and GGBS generated annually is still disposed in landfills or disposal sites (Kumar et al. 2005; Naganathan et al. 2012; Domtar 2015). Massive quantities of natural sand is being used all over the world for various engineering applications due to which natural sand deposits are being depleted and causing serious threat to environment as well as on society. Therefore considering for an alternative to natural sand is very much necessary. The cheapest alternate to natural sand is quarry sand/quarry dust. Another substitute to natural sand is processed slag sand (PSS Sand), can be used as fine aggregate in the development of Controlled Low Strength Materials.

Wu et al. (2016) conducted laboratory studies to evaluate the feasibility and applicability of incorporating disposed paper sludge from paper mills into CLSM. In their study fly ash is effectively used by limiting the amount of cement which results in the flow ability of the mixtures and concluded that strength of CLSM was reduced when higher content of fly ash was incorporated to replace cement.

Alizadeh et al. (2014) developed optimized CLSM mixture to be used as structural fill for construction of a bridge abutment. Series of CLSM mixtures were developed and tested in the laboratory for engineering properties such as flow ability, density, compressive strength and stress–strain behavior. Since it was an important area of concern in design of the CLSM bridge abutment, the bond strength performance of the CLSM to steel anchors were studied under numerical simulation of pullout tests. They concluded that CLSM mixture with higher compressive strength resulted in higher bond strength and more brittle slippage and bond strength decreases with increase in bar size and embedment length was observed in pull out test.

Le et al. (2016) investigated the fresh and hardening properties of a soil-based controlled low-strength material (CLSM) containing stainless steel reducing slag (SSRS) as a cementitious material for trench-fill. Four sets of cement substitution (i.e. 0%, 10%, 20%, and 30%) and three sets of binder content (80-, 100-, and 130 kg/m3) were used for this experimental work. Fresh and hardening properties of the recommended CLSM were studied in the laboratory. Flow ability of the proposed CLSM was better as the percentage of SSRS was increased. They concluded that the higher SSRS replacement will reduce both compressive strength and elastic modulus and the proposed CLSM could be employed for eco- friendly trench fills.

Hwang et al. (2017) evaluated the properties of CLSM, prepared using a mixture of class-F fly ash (FA) and ground granulated blast-furnace slag (GGBFS) with ratios (0%, 10%, and 20% as calculated by combined FA and GGBFS weight) of waste water treatment sludge (WTS). A sodium hydroxide (NaOH) solution was used to activate these powder materials in the mixture in order to produce the alkali-activated CLSM. The effects of adding WTS on the properties of both fresh and hardened CLSM are evaluated along with toxicity characteristic leaching procedure (TCLP) tests, further Scanning electron microscope (SEM), Energy dispersive spectrometer (EDS) and X-ray diffraction (XRD) was used to examine the micro structural properties of the CLSM. They observed and noted that adding WTS to the CLSM samples reduces workability and increases compressive strength. Further the liquid-to-solid (L/S) ratio and the alkali equivalent (AE) affects the properties of the CLSM. Finally, the proposed CLSM sample with an L/S ratio of 0.9, WTS of 10%, and AE of 9% are found to very much satisfactory.

Bassani et al. (2015) carried out an investigation on long-term stiffness characterization of Controlled Low-Strength Materials (CLSMs) for pavement in replacement with granular fill materials. Three alternative CLSM mixtures, two with OPC and a third one with ultra-rapid sulpho-aluminate cement, were tested. Two different sample aspect ratios were considered and the samples were subjected to saturation, loading time and repetition. From the results of their experimental investigation it is clear that proposed CLSMs are not sensitive to variations of loading frequency and to water saturation, but sensitive to sample aspect ratio, and there is an increase in stiffness under repeated load tri-axial testing and a low permanent strain accumulation. Finally, there is an increase in resilient modulus when the deviatoric stress was increases.

The literature review clearly indicates the absence of field controlled procedures based on model development for bulk utilization of different marginal materials and variation in batch quality. Reliability for the required specification is not guaranteed. Also use of GGBS + Class C fly ash in cementitious matrix composites is limited and needs further exploratory research. Modelling of sustainable cementitious matrix composites is scarce. Study of sustainable materials for field application has been minimal and requires use of scientific approaches and modelling for effective bulk utilization leading to carbon footprint mitigation. Development of models for CLSM using bulk produced waste by-products as zero energy carbon efficient materials for reliable sustainable construction practice.

The main objective of the present study is to obtain novel CLSM mixes by utilizing industrial wastes (Ling et al. 2018) such as Class C fly ash, GGBS, quarry dust/sand and processed slag sand; and to develop phenomenological models which can be effectively utilized by the practicing engineers.

2 Materials and Methods

2.1 Materials Used

Binders

1. Cement (C): 53 grade; density = 3.14 g/cm^3
2. GGBS (G): density = 2.91 g/cm^3
3. Class-C fly ash (F): density = 2.38 g/cm^3

Fine Aggregate

1. Quarry Sand: density = 2.61 g/cm^3
2. Processed Slag (PS) Sand: density = 1.3 g/cm^3

Type of Curing

1. Air curing at room temperature

Tests Conducted

1. Fresh state: flow test and Marsh cone - for workability
2. Hardened state: strain-controlled-unconfined-compression test on cylindrical Specimens

Size of Specimen

1. Diameter = 40 mm; height = 80 mm

Variable Parameters

1. Cement + Class C fly ash + GGBS: PS sand + Quarry Sand (1 binder: 1 aggregate)
2. Binder ratio (F + G/C): 10, 6, 5, 4, 3 and 2
3. Water content % = 35, 38, 41 and 44
4. Compressive Strength tests = at the age of 7, 28, 56 and 90 days.

Table 1 illustrates the material calculations. For each dry mix series 20 specimens were cast.

2.2 Methods

Flowability

One of the notable advantages of CLSM as compared to the conventional backfill materials is its consistency when it is fresh. Because of the good flowability property of CLSM, it can be very easily placed compared to conventional backfill materials and it does not require any compaction or vibration (Du 2008). This will reduce labor cost, and also the duration of the construction (Du et al. 2002). It is necessary to develop CLSM mixture which does not segregate during placing in the trenches and when filling embankments backfill. The consistency and workability of CLSM is determined by conducting spread flow experiment using acrylic cylindrical mould. The dimensions of acrylic cylindrical mould are 150 mm high with 75 mm diameter. In this test coarse aggregates are not used to prepare the CLSM mix instead fine aggregates in the form of processed slag sand and quarry sand is used along with other industrial wastes such as Class C fly ash and GGBFS. Most of the materials used in preparing the CLSM are marginal/waste materials. The cylinder is placed on a smooth base plate made of glass or brass circular plate of about one meter diameter and filled with fresh CLSM.

Table 1. Material calculations

Series ID	w %	Cement (g)	Class C Fly ash (g)	GGBS (g)	PS Sand (g)	Quarry dust/sand (g)	Total dry mix (g)	Water (ml)
FG1	35	100	200	800	550	550	2200	770
FG2	38	100	200	800	550	550	2200	836
FG3	41	100	200	800	550	550	2200	902
FG4	44	100	200	800	550	550	2200	968
FG5	35	150	200	800	575	575	2300	805
FG6	38	150	200	800	575	575	2300	874
FG7	41	150	200	800	575	575	2300	943
FG8	44	150	200	800	575	575	2300	1012
FG9	35	200	200	800	600	600	2400	840
FG10	38	200	200	800	600	600	2400	912
FG11	41	200	200	800	600	600	2400	984
FG12	44	200	200	800	600	600	2400	1056
FG13	35	250	200	800	625	625	2500	875
FG14	38	250	200	800	625	625	2500	950
FG15	41	250	200	800	625	625	2500	1025
FG16	44	250	200	800	625	625	2500	1100
FG17	35	300	200	800	650	650	2600	910
FG18	38	300	200	800	650	650	2600	988
FG19	41	300	200	800	650	650	2600	1066
FG20	44	300	200	800	650	650	2600	1144
FG21	35	350	200	800	675	675	2700	945
FG22	38	350	200	800	675	675	2700	1026
FG23	41	350	200	800	675	675	2700	1107
FG24	44	350	200	800	675	675	2700	1188

The cylinder is filled completely with prepared mixture of CLSM and filled without any air entrapment and lifted. Due to its own weight the mix spreads whose diameter in mutually perpendicular directions is measured and averaged. The relative flow area (RFA) is calculated from the relation.

$$\mathrm{RFA} = \left(\mathrm{D}^2 - 75^2\right) / 75^2 = [\mathrm{D}/75]^2 - 1 \text{ (Raghavendra and Udayashankar 2015)} \quad (1)$$

Where D = average diameter of mix spread, in mm.

Setting time of fresh CLSM was evaluated as per ASTM C403 (Raghavendra et al. 2016a, b). Settlement was evaluated considering reduction in height at increasing ages with respect to initial fresh CLSM specimen height.

Phenomenological Models

The main aim of the study is to make use of the industrial by-products and other wastes as a constituent material in the preparation of CLSM, to generate a set of experimental data, analyze them, and formulate a phenomenological model, i.e., a kind of ready-to-use formula. The use of ready to use equations would be to proportion CLSM mixes for desired strength (at specified age and binder/water ratio) and flow (at specified water content). As mentioned in literature (Raghavendra and Udayashankar 2014); "Du et al. (2002) made efforts to develop predictive models for the compressive strength of CLSM; various models and statistical approaches were considered but no single model was found to work well for the entire range of materials and mixture proportions. The various combinations of materials involved in the generation of CLSM mixes would involve multiple iterations to achieve desired flow and strength. With the help of phenomenological model it is required to make a single trial and determine its flow and strength values, respectively. Further this trial mix value would enable us to predict flow and strength of other mixes within no time and without much time-consuming effort. The two basic principles underlining any development of procedures to combine different ingredients are Abrams' law and Lyse's rule. Further these basic laws have been generalized to encompass different sets of ingredients (Nagaraj and Banu 1996). These two basic principles have already been examined in detail for proportioning the normal concrete mixes. It is not clear as yet what modifications are needed if it has to be valid for cement-based controlled low-strength materials mainly because of the low strength range and high range of water/cementitious ratios involved. High values of these ratios are to be employed to lower the strength levels to be mobilized. Even with regard to the workability the range to be covered is that for self-leveling and self compacting state of the mixes. The present investigation aims to generalize the two basic principles namely Abrams' law and Lyse's rule for proportioning the cement-based controlled low strength materials".

3 Results and Discussions

Flow and Strength

Averaged spread flow diameter values are illustrated in Fig. 1. The flow diameters ranged from 21.75 cm to 55.5 cm. The flow values were assessed in terms of RFA and found that most of the values were above 15. In literature it is prescribed that 5–15 RFA is ideal considering self-flow and levelling properties of CLSM (Udayashankar and Raghavendra 2014). The higher water contents (41% and 44%) chosen; resulted in segregation and bleeding behavior. Predictive flow phenomenological models were developed by normalizing flow RFA values with respect to 38% water content considering 5-15 RFA range. FG13 to FG16 series RFA values are used to validate flow models and hence were not included to develop flow predictive models. The generalized flow model for FG series is "$\{RFA/(RFA@w = 38\%)\} = 0.2377w - 7.8293$". The flow model includes excess-flow values of higher water contents and hence its prediction was not acceptable at those water contents. However, flow prediction values at lower water contents were encouraging and recommended for practical use.

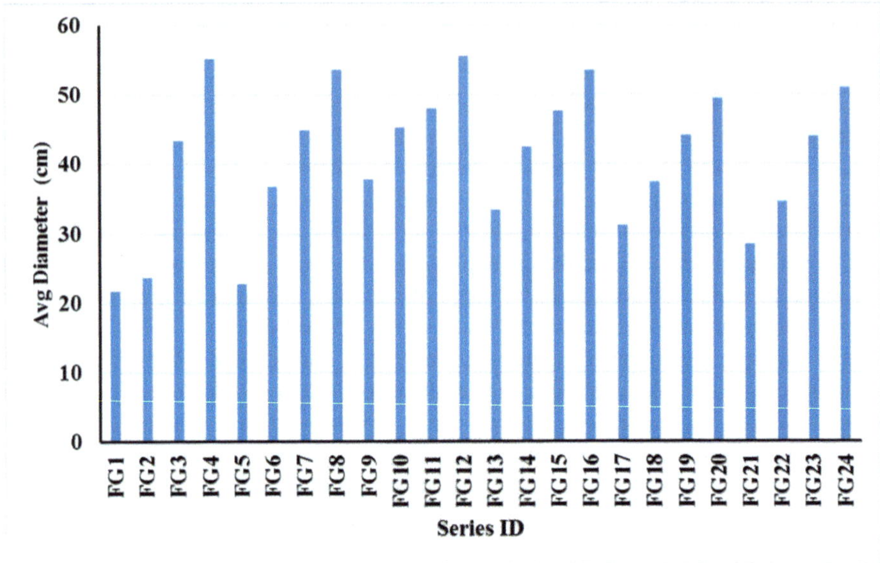

Fig. 1. Average flow results

Averaged Marsh flow values ranged from 16 s to 35 s and zero flow was recorded at lower water contents (35% and 38%).

Averaged compressive strength values are illustrated in Fig. 2. The 7 day and 28 day strength values ranged from 0.12 MPa to 0.91 MPa and 0.16 MPa to 1.29 MPa, respectively. The 56 day and 90 day strength values ranged from 0.102 MPa to 1.35 MPa and 0.086 MPa to 1.442 MPa, respectively. Higher strengths for any constant dry mix proportion were observed for lowest water contents and vice versa (Raghavendra et al. 2016a, b). Due to pozzolanic action of the binder ingredients, strength values increased with increase in age. However, few dry mix proportions recorded almost same strength values at both 56 day and 90 day age. Predictive strength phenomenological models were developed by normalizing strength values with respect to 1.3 B/w ratio i.e. 38% water content considering 5-15 RFA range. FG13 to FG16 series strength values are used to validate strength models and hence were not included to develop strength predictive models. The generalized strength (S) models for FG series at 7, 28, 56 and 90 day age are *"{S/(S@B/w = 1.3)} = 1.619(B/w) – 1.0559"*; *"{S/(S@B/w = 1.3)} = 1.0543(B/w) - 0.342"*; *"{S/(S@B/w = 1.3)} = 0.5108 (B/w) + 0.3739"*; *"{S/(S@B/w = 1.3)} = 0.8246(B/w) - 0.0635"*; respectively. Figure 3 illustrates the validation of predicted strength values for FG13 to FG16 series using the above mentioned strength phenomenological models. The strength prediction values were well within the limits required for engineering decisions and hence recommended for practical use.

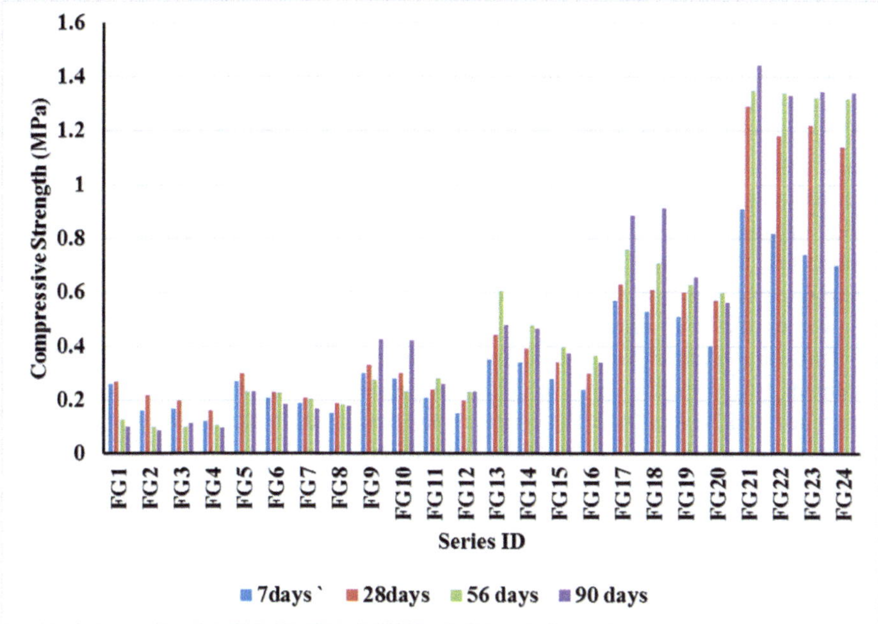

Fig. 2. Average compressive strength results

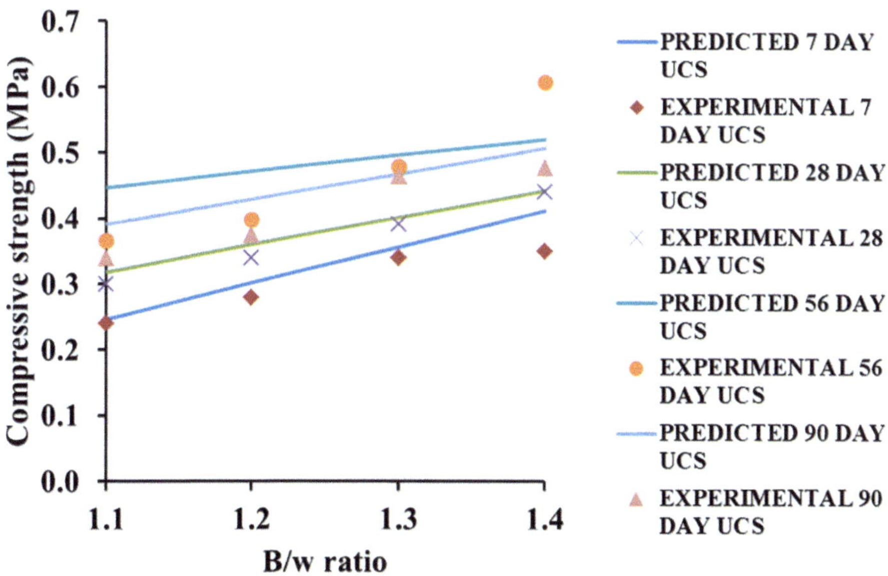

Fig. 3. Validation of strength models (FG13 TO FG16 SERIES)

Settlement and Setting Time

Averaged setting time values are illustrated in Fig. 4. The setting time values ranged from 20 min to 40 min. This indicates that all the dry mix proportions are such that it results in a fast setting /quick setting CLSM mixture. However, lower setting times were observed for higher water contents and vice versa. It may be noted that the flow values were not encouraging at higher water contents, and considering the time required for a particular activity, one may however choose an excess flow (>15 RFA) CLSM mixture if the setting time requirement is around 40 min. Hence it is recommended that these novel FG series CLSM mixes are highly suitable for applications requiring early repair of infrastructures restricting traffic flow for just an hour. Settlement values of all FG series mixes ranged from 1 to 2 mm up to 7 days age. At 28 days age it was in the range 2 to 3 mm. And thereafter at the ages of 56 days and 90 days, settlement was constant i.e. 2 to 3 mm. It may be noted that these values include subsidence and hence these FG series mixes performed better even at higher water content mixes resulting in excess flow (>15 RFA).

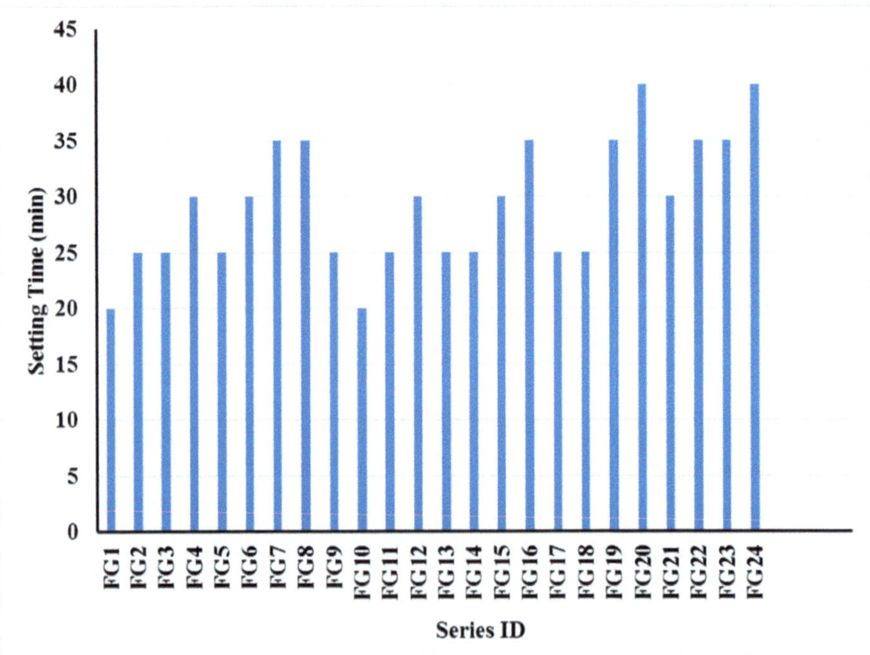

Fig. 4. Average setting time results

4 Conclusions

The following conclusions are made based on the above discussed results

1. FG series CLSM mixes incorporating Class C fly ash and GGBS with PS sand and Quarry sand, resulted in quick setting or fast setting CLSM mix.
2. Irrespective of the high water contents chosen, all FG series mixes recorded setting times in the range of 20–40 min.
3. FG series mixes are more suitable for quick repair activities requiring lesser time of traffic flow restriction.
4. Most of the FG series CLSM mixes are re-excavatable since strength values are within the recommendations of ACI-229R.
5. Maximum of 3 mm settlement was recorded for highest water content CLSM mix, and hence is well the limits prescribed by ACI-229R.
6. The Strength phenomenological models predicted strength values within reasonable engineering accuracy levels.
7. The flow phenomenological models predicted acceptable flow values only for lesser water contents (5-15 RFA (Udayashankar and Raghavendra 2014))
8. Sustainable CLSM mixture is developed using marginal and waste materials in large quantities essentially for quick infrastructure repair works.

Acknowledgments. A portion of this research was performed using facilities at Department of Civil Engineering, R. V. College of Engineering, Bengaluru; and Civil-Aid Technoclinic Pvt. Ltd. (A Bureau Veritas Group Company), Bengaluru.

References

Hwang, C.-L., Chiang, C.-H., Huynh, T.-P., Vo, D.-H., Jhang, B.-J., Ng, S.-H.: Properties of alkali-activated controlled low-strength material produced with waste water treatment sludge, fly ash, and slag. Constr. Build. Mater. **135**, 459–471 (2017)

Domtar: Sustainability Report, Domtar Corporation (2015)

Du, L., Folliard, K.J., Trejo, D.: Effects of constituent materials and quantities on water demand and compressive strength of controlled low-strength material. J. Mater. Civ. Eng. 485–495 (2002). https://doi.org/10.1061/(ASCE)0899-1561(2002)14:6(485)

Du, L.: Laboratory investigations of controlled low strength material. Ph.D. dissertation, University of Texas at Austin, Austin, TX (2008)

Le, D.-H., Nguyen, K.-H.: An assessment of eco-friendly controlled low-strength material. Procedia Eng. **142**, 260–267 (2016)

Wu, H., Huang, B., Shu, X., Yin, J.: Utilization of solid wastes/by-products from paper mills in controlled low strength material (CLSM). Constr. Build. Mater. **118**, 155–163 (2016). https://doi.org/10.1016/j.conbuildmat.2016.05.005

Kumar, V., Mathur, M., Sinha, S.S., Dhatrak, S.: Fly ash Environmental Saviour. Report Submitted to Fly ash Utilization Programme (FAUP) to TIFAC, DST, New Delhi, India (2005)

Bassani, M., Khosravifar, S., Goulias, D.G., Schwartz, C.W.: Long-term resilient and permanent deformation behaviour of controlled low-strength materials for pavement applications. Transp. Geotech. **2**, 108–118 (2015)

Naganathan, S., Razak, H.A., Hamid, S.N.A.: Properties of controlled low-strength material made using industrial waste incineration bottom ash and quarry dust. Mater. Des. **33**, 56–63 (2012). http://dx.doi.org/10.1016/j.matdes.2011.07.014

Nagaraj, T.S., Banu, Z.: Generalization of Abram's law. Cem. Concr. Res. **26**(6), 933–942 (1996)

Raghavendra, T., Udayashankar, B.C.: Flow and strength characteristics of CLSM using ground granulated blast furnace slag. J. Mater. Civ. Eng., (ASCE) **26**(9), 04014050 (2014). https://doi.org/10.1061/(asce)mt.1943-5533.0000927

Raghavendra, T., Udayashankar, B.C.: Engineering properties of controlled low strength materials using flyash and waste gypsum wall boards. Constr. Build. Mater. **101**(Part 1), 548–557 (2015). https://doi.org/10.1016/j.conbuildmat.2015.10.070

Raghavendra, T., Sunil, M., Udayashankar, B.C.: Controlled low-strength materials using bagasse ash and fly ash. ACI Mater. J. **113**(4), 447–457 (2016a). https://doi.org/10.14359/51688987

Raghavendra, T., Siddanagouda, Y.H., Jawad, F., Adarsha, C.Y., Udayashankar, B.C.: Performance of ternary binder blend containing cement, waste gypsum wall boards and blast furnace slag in CLSM. Procedia Eng. **145**, 104–111 (2016b). https://doi.org/10.1016/j.proeng.2016.04.027

Ling, T.-C., Kaliyavaradhana, S.K., Poon, C.S.: Global perspective on application of controlled low-strength material (CLSM) for trench backfilling – An overview. Constr. Build. Mater. **158**, 535–548 (2018). https://doi.org/10.1016/j.conbuildmat.2017.10.050

Udayashankar, B.C., Raghavendra, T.: Proportioning controlled low strength materials using fly ash and ground granulated blast furnace slag. In: Ohji, T., Matyáš, J., Manjooran, N.J., Pickrell, G., Jitianu, A. (eds.) Advances in Materials Science for Environmental and Energy Technologies III: Ceramic Transactions, vol. 250. Wiley, Hoboken (2014). https://doi.org/10.1002/9781118996652.ch2

Alizadeh, V., Helwany, S., Ghorbanpoor, A., Sobolev, K.: Design and application of controlled low strength materials as a structural fill. Constr. Build. Mater. **53**, 425–431 (2014)

New Theory of Soil Response to a High Energy Impact and Its Environmental Consideration

Khelalfa Houssam[1,2(✉)]

[1] Department of Civil Engineering, University of Mohammed Seddik Ben Yahia, Jijel, Algeria
khelalfahoussam@gmail.com
[2] Department of Civil, Geotechnical and Coastal Engineering, K.E.C Laboratory, Jijel, Algeria

Abstract. Dynamic compaction (DC) is one of the techniques of soil improvement. The impact creates body and surface waves that propagate in the middle of the ground. Soil response to high energy impact during dynamic compaction is very complex. This may be the reason for the slow progress in the development of a rational method for dynamic compaction analysis. However; very few theoretical models are available, which prevents us from understanding all the side effects resulting from it. From US; a case study about the adoption of the dynamic compaction technique with high energy was presented, to facilitate the analysis of the results that cannot be explained by the theory of Lukas, 1986. From the consequences we have found; We proposed a theory (September, 2017 by H. Khelalfa) of soil response to a high energy impact during dynamic compaction. Our theory is based on that there is a reflection of part of the energy received by the compacted soil matrix, implied by the distribution of the energy (E) into two parts, the effective energy (E′) that the sole responsible for soil compaction, and the lost energy (Ep) that the responsible of the neighbouring resulting vibrations. This gives us a wider field for interpreting the results associated with dynamic compaction. It is also observed that the equations obtained through our theory are enter in order to facilitate the calculation of certain parameters. If we take the equations of our theory, we can easily obtain the graphs of the prediction of the level of vibration transmitted by the soil which will influence the environment. The graph of the energy response is divided into four (04) parts (very dangerous, dangerous, limited non-dangerous and non-dangerous vibration); to give us an approach to estimate the efficiency of dynamic compaction and its effects (vibrations).

Keywords: Dynamic compaction · Energy · Vibrations · Environment New theory

1 Introduction

In recent years, the Algerian authorities have been considering the development of container traffic at the Djen-Djen port of Jijel, both for the local market and for transshipment between the collecting vessels. The port valuation study led to the development of a new container terminal, to be built in the central part of the port.

H. Ameen and T. Sorour (Eds.): GeoMEast 2018, SUCI, pp. 87–104, 2019.
https://doi.org/10.1007/978-3-030-01929-7_7

In order to ensure the time and quality of the project, a workshop of dimensions 11,400 m^2 for manufacturing caissons and 40,945 m^3 of embankment of the dredged sand is essential (Fig. 1).

Fig. 1. Photographies of steps of backfilling of the caisson's workshop in the Djen-Djen port of Jijel, Algeria.

Geotechnician is interested in the soil as it is the main element of the context in which the stability of a structure will be conceived [1]. It was therefore quickly considered to improve the mechanical characteristics of the backfill to increase their bearing capacity and eliminate settlements and risks of liquefaction [2]. Soil improvement methods are one of the tools available to the engineer to solve stability problems or deformations that encounter when developing a project. A large number of processes exist; on the one hand, and on the other hand, mechanical compaction processes, mainly implemented on hydraulic backfill [3].

2 Land Reclamation

Hydraulic embankments are often used to recover land for large infrastructure projects such as airports, ports, industrial areas. Given the fundamental importance of hydraulic backfill for infrastructure projects, a need for quality of material borrowed and construction methods are crucial for the quality of the final product, which will require specific performance requirements. Land reclamation is generally defined as the process of creating new land by raising the elevation of a seabed, or other land at low altitude. it can be carried out by a movement of dry earth, also by hydraulic filling [4]. Some possible failure modes in the embankment body and different failure modes need to be analyzed. From the perspective of the foundation, this can pose a significant risk of partial or complete liquefaction and consequently reducing soil resistance. Global Failure Stability Analysis provides suggestions for improvement methods to be performed [5].

3 Improvement of Embankments

For big projects, where soil mechanics play an important role, the corresponding cost and construction time seem unpredictable. The geotechnical engineer should be involved in the early stages of a project and be better able to convince the client of the need for a good soil survey. Surveys of the sites are determined by their needs and their impact, specific requirements, the specific site and its environment, as well as Economic [6]. During the feasibility study of a project, the use of soil treatment methods requires knowledge of their performance and limitations. A question then arises: how to represent in an easily usable way the fields of application of each process? We have chosen to represent the ability of a method to treat a soil according to its granulometry. It has the advantage of only using identification criteria obtained by simple laboratory measurements (Fig. 2). However, it is clear that the other identification factors and the mechanical parameters of soils are to be taken into account in the precise definition of the treatment of each concrete case [7].

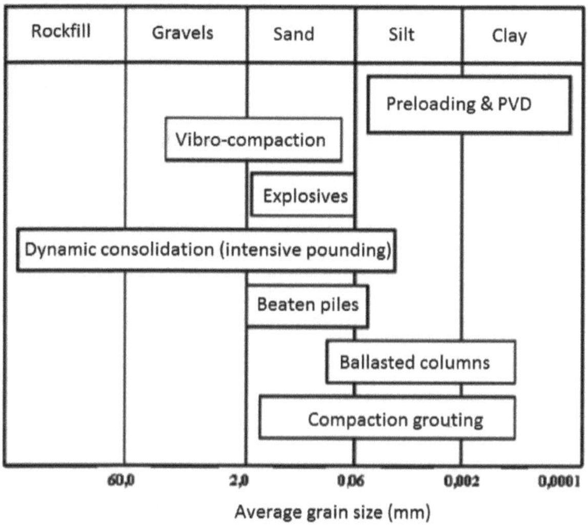

Fig. 2. Distribution of treatment methods according to granulometry.

4 Dynamic Compaction (DC)

Dynamic compaction (DC) is one of the techniques of soil improvement. It depends on the rearrangement of the soil particles using the dynamic energy produced by falling a weight (tamper) from a certain height [8], (Figs. 3 and 4). The concept of this technique is to improve the mechanical properties of the soil by transmitting high energy impacts on loose soils that initially have low load capacity and high compressibility potentials. Impact creates body and surface waves that propagate in the middle of the ground [8].

Fig. 3. Free fall of a tamper of 20 ton at the of Djen-Djen port in Jijel, Algeria.

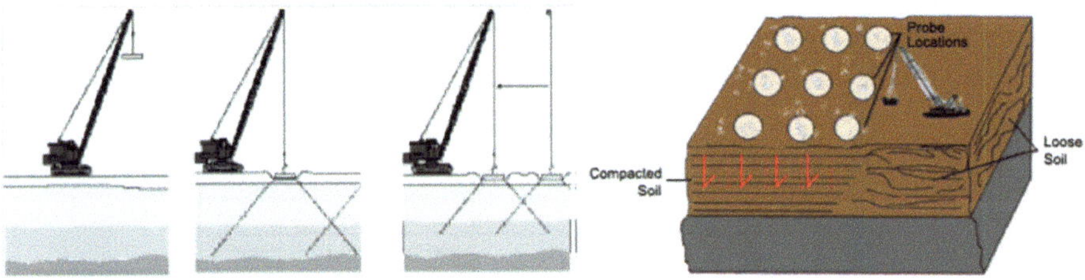

Fig. 4. Application of dynamic compaction in phases [8].

Soil response during dynamic compaction treatment varies with soil type and energy supply. In unsaturated soils, the waves move the soil grains and rearrange them into a denser configuration. In saturated soils, the soil is liquefied and the grains rearranged in a more compact state. In both cases, the decrease of the voids and the increase of the internal granular contact will lead directly to the improvement of the soil properties [7].

5 Execution of Dynamic Compaction in the DjenDjen Port, Jijel, Algeria

5.1 Study of the Soil to Compacted

The laboratory tests usually for the design of embankments are those for the determination of the index properties (granulometry analysis, liquid and plasticity limits, grain specific gravity). Granulometry has the advantage of only using identification criteria obtained by simple laboratory measurements, which allows the availability of time and project costs. In order to apply dynamic compaction in an efficient manner, it is necessary to study and determine the properties of the terrain along the required depth (Figs. 5 and 6), to be classified in the desired area (Fig. 7). Lukas [10] has classified the soil into three improvement zones based on soil classification; According to laboratory tests, the project site belongs to zone-1 (red line inside Fig. 7) and dynamic compaction is therefore effective:

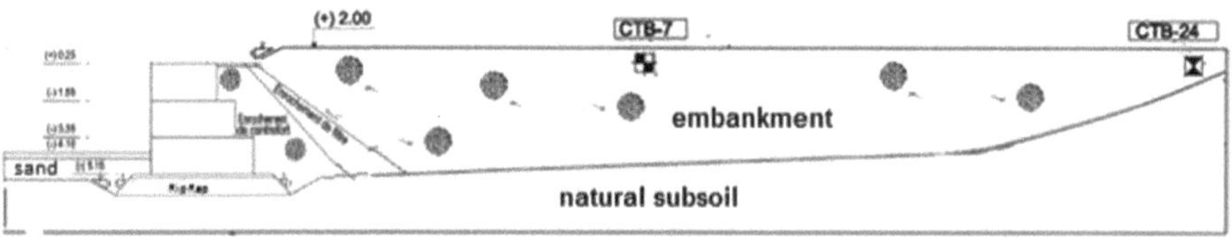

Fig. 5. Cross-section of the workshop embankment before DC [9].

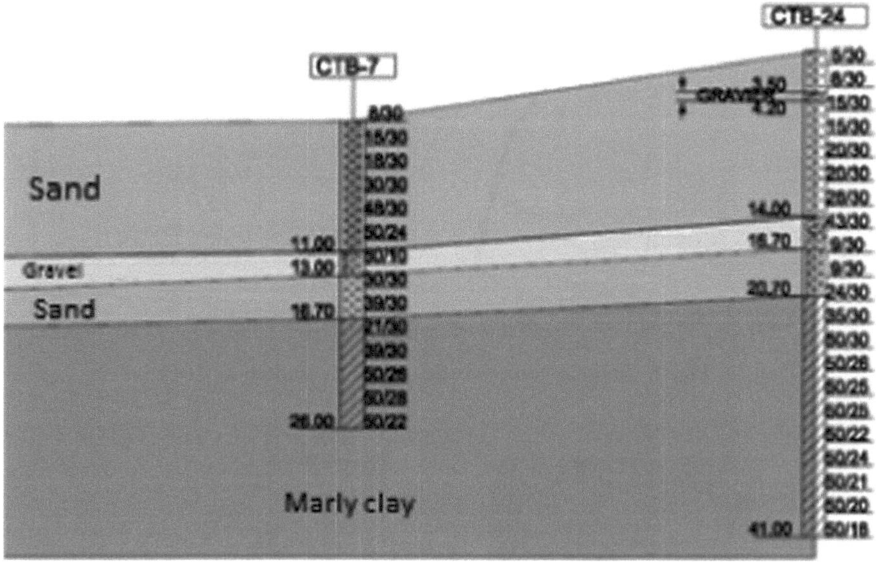

Fig. 6. Lithological section of the natural ground (CTB-7, BTC-24) below the workshop embankments.

- Zone-1: ideal floors for applying dynamic compaction,
- Zone-2: Possible floors for applying dynamic compaction,
- Zone-3: impossible floors for applying dynamic compaction.

5.2 Test Board

The test board is carried out with different distance (Fig. 8) and various numbers of strokes (13 to 26) to determine the appropriate mesh size for good soil improvement which ensures the stability of the embankment of caisson's workshop. The height of fall (H = 17 m) and the weight of the tamper (M = 20 tons) used for dynamic consolidation are determined by the capacity of the fall device and the depth of improvement. The possible depth to be improved related to the blow energy by a stroke. After each stroke, the settlements must be checked by an optical level. Once made the first series, it is necessary to check if it appears water at the bottom of the crater. After the first two series of strokes we must be to perform the third series.

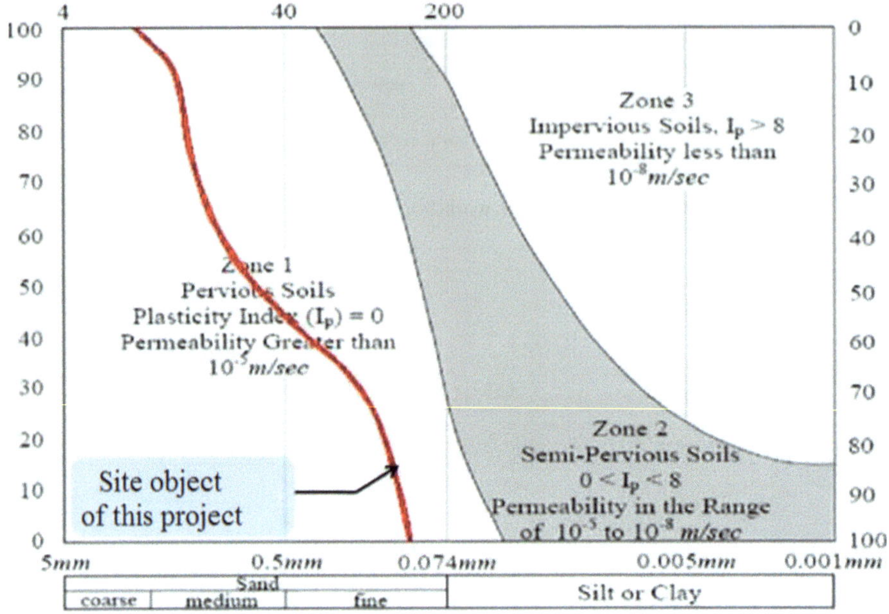

Fig. 7. Soil categorization for dynamic compaction [10].

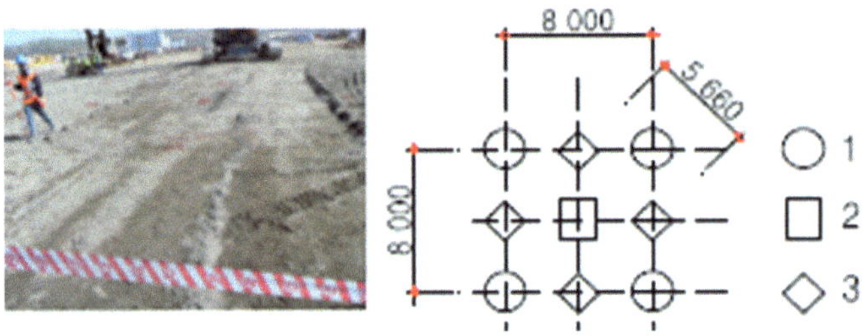

Fig. 8. The square of the test board.

5.2.1 1st Test Board (PE-1; CTB-24)

The number blow of reasonable compaction is obtained by taking into account the penetration depth by compacting during test. After calculating the number blow of compaction taking into account the penetration depth by compaction during the test. The reasonable number blow of compaction is 11 in the 1st stage, and 12 in the 2nd stage, 10 in the 3rd stage. After calculating the compression ratio according to the number blows of dynamic compaction (1st stage), the compression ratio is 1.536 m. In the second stage of compaction, the compression ratio is 1.28 m. In the 3rd compaction stage, the compression ratio is 1.034 m (Fig. 9). He compression ratio decreases since the number of compaction increases in steps, because the space between particles decreases and the density of the soil increases with consolidation.

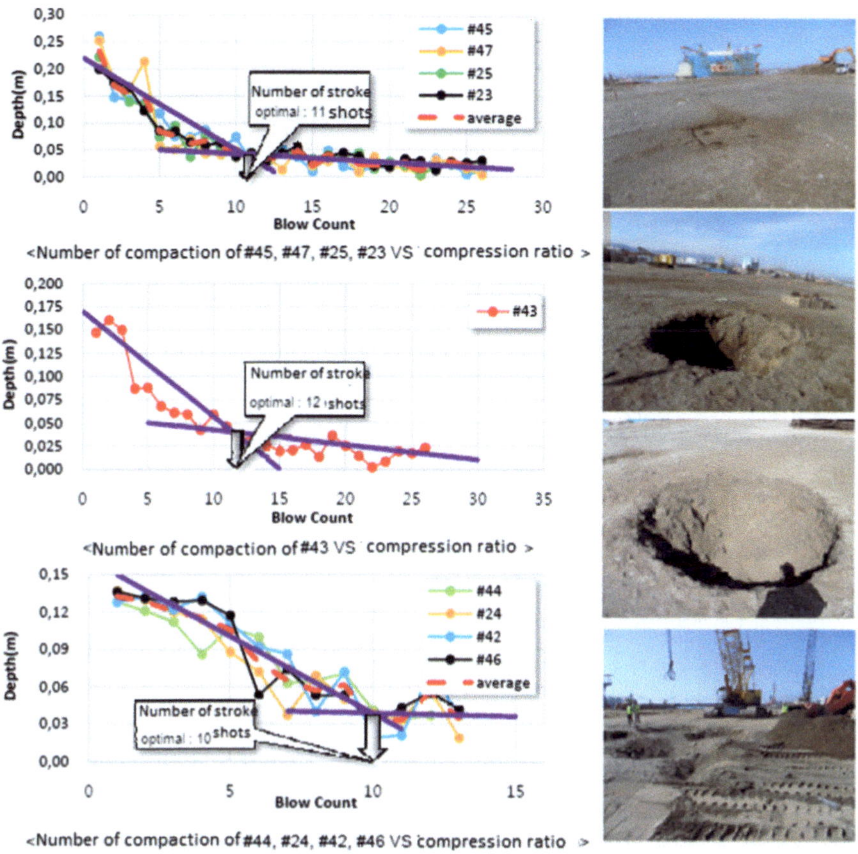

Fig. 9. Compaction in steps vs compression ratio (accumulated) of PE-1.

5.2.2 2nd Test Board (PE-2; CTB-7)

If we adopt the criterion of the completion of a settlement of 2.5 cm (1 inch) to stop the compaction, we can see in the graph (Fig. 10) that in the first series, to obtain a settlement of 2.5 cm (1 inch), it is necessary to wait until the 19th blows, on the other hand, in the second series it is necessary to arrive at the 14th blows. Compared to the third series (green line) we can observe that the settlements for each stroke are lower than those of the other stages and that it is possible to stop the third series from the 10th blows. If we analyze the graphs of the settlements of the first and second series, we can see that, after the first series, the ground is consolidated, and consequently, the settlement of the second series is less.

5.3 Compacting Workshop Embankment

After analyzing the results of the test board, it is requested to initiate the dynamic compaction of the necessary portion of the workshop according to the intended plane (Fig. 11).

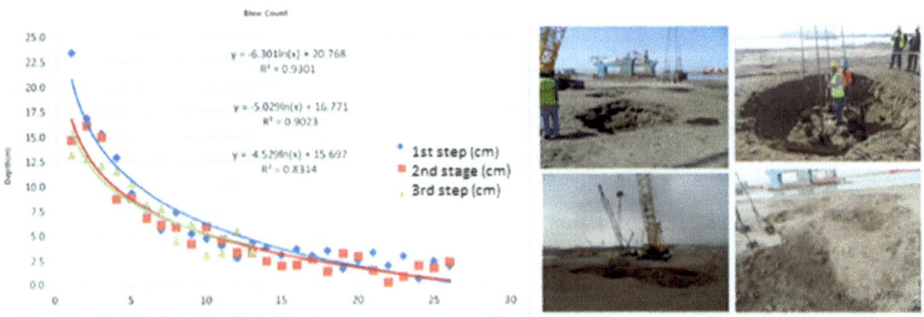

Fig. 10. Compaction in steps vs compression ratio (accumulated) of PE-2.

Fig. 11. DC improvement section of the workshop.

During the compaction of the second (2nd) line of the caisson's foundation station we were surprised? !! disaster !? The appearance of cracks (The presence of cracks in the floor and beams of the building) in the Harbor Authority building, which is about 108 meters from the site of DC operation (Fig. 12). It has been proposed to make a trench as a solution in the service stations, to prevent the transmission of vibrations to the building. For tracking if there is a progression of cracks after the execution of the proposed solution, it was asked to make a plaster paste on each crack (la Fig. 13). After the completion of all precautions, compaction was continued, where no crack development was observed later.

The remaining section of the workshop continued to be compacted, leaving a precautionary distance of 7 m (Fig. 11) away from the embankment retaining blocks, where the compaction energy was reduced to preserve the stability of the blocks. Despite having compacted without reducing the energy of the compacted in the line

Fig. 12. Photographs of cracks in the port authority resulting from DC.

Fig. 13. Photographs of the trench and the plaster paste.

before last, although it is a very close distance; we did not notice any impact on the stability of the blocks.

6 New Theory

The response of the soil to a high energy impact during dynamic compaction is very complex. This may be the reason for the slow progress in the development of a rational method for dynamic compaction analysis. Very few theoretical models are available [11]. From US; Considering the consequences (Fig. 12) of dynamic compaction of the workshop embankment at the Djen-djen port, Algeria, another theory [12] (Fig. 14b) of the soil response to high energy impact during dynamic compaction has been proposed to facilitate analysis of results (Fig. 12) that cannot be explained by Lukas's theory, 1986 [9] (Fig. 14a).

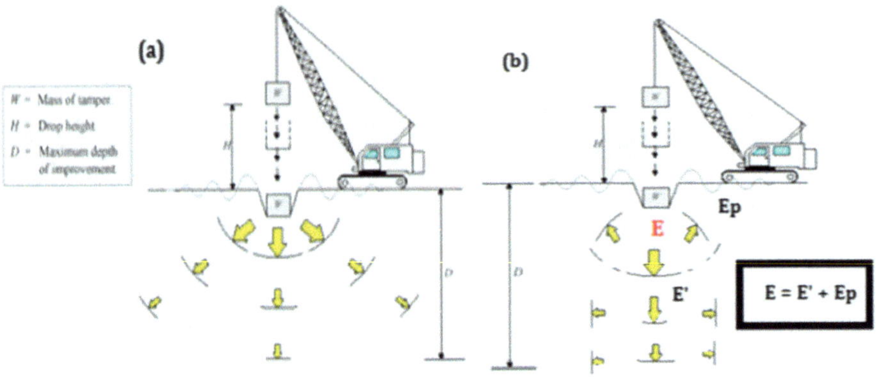

Fig. 14. Dynamic compaction after, (a) Lukas [10], (b) Khelalfa [12].

As mentioned in Eq. 1, Ménard is defined the energy (E) by the multiplication of the mass of the tamper (M) and the height of fall (H);

$$E = M.H \tag{1}$$

Our theory is based on the existence of a **reflection** of a part of the energy received by the matrix of the soil to be compacted (Fig. 14b), which implies a distribution of energy in two parts (Eq. 2), the **effective energy** (E′), which is solely responsible for soil compaction, and the **lost energy** (Ep) responsible for neighboring resultant vibrations:

$$E = E^{'} + Ep \tag{2}$$

6.1 DC Mechanism

The physical displacement of the particles and, to a lesser extent, the low-frequency excitation caused by stress pulses will then rearrange the particles in a denser state [13], and reduce the void ratio and increase the relative density to provide improved characteristics of bearing capacity and settlements [14, 15]. Therefore, it can be considered that the energy is a function of the void ratio (e) Eq. 3:

$$E = f(e) \tag{3}$$

When granular materials extend below the water table (groundwater), a high proportion of the dynamic impulse is transferred to the porous water which, after an appropriate number of surface impacts, finally rises up to a level sufficient to induce liquefaction [14, 15]. In this case we will consider that the energy is a function of degree of saturation (S) instead of the void ratio Eq. 4:

$$E = f(S) \tag{4}$$

6.2 Case of Dry/Wet Soils

In dry/wet granular soils, the physical displacement of particles and therefore the collapse of the soil skeleton (pothole) is the main mechanism of compaction. During the DC process, high energy waves propagate through the ground, Thereby, each soil element experiences strong confining pressure for a very short period of time. The mechanically agitated granular materials are characterized by slow relaxation dynamics, resulting from the rearrangement of constituent grains in the volume in which they are confined [16]. The collective motion of the sliding particles and rearrangements of the grains, mainly by sliding is the main process responsible for the relaxation of the constraint [17]. Given the Eq. 3, we deduce the Eq. 5:

$$E^{'} = E.e \tag{5}$$

A large damping (amortization) sufficiently slows down the system so that the grains have sufficient time to develop substantial tangential displacements, which in turn generate sufficiently high shear forces to cause the sliding of the grains. On the other hand, it is also known that aging occurs at contacts between particles [14, 15], which manifests itself as an increase in the coefficient of friction between grains as a function of time. when the damping (attenuation) in the system exceeds a critical value, there is a slow increase in the number of sliding particles as a function of time, this slow strengthening leads to constraints relaxation [17]. Given the Eq. 3, we deduce the Eq. 6:

$$Ep = E.(1 - e) \tag{6}$$

The overall process of dynamic compaction in dry/wet granular soils can be separated into three major zones (Fig. 15) [12]:

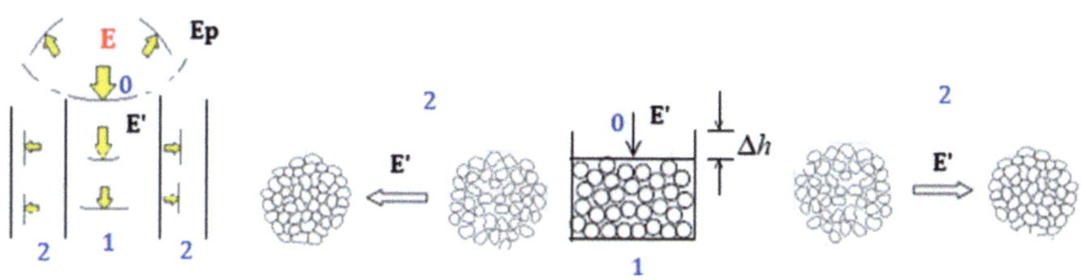

Fig. 15. Overall mechanism of dynamic compaction in dry/wet granular soils [12].

- ZONE (0): the collision zone; between the tamper and the ground; where the total deformation of the soil matrix, the crushing of the grains during shocks, resulting in a crater and a discharge of soil.
- ZONE (1): the sliding zone; under the effect of effective energy, the grains move by sliding to a perfectly dense state.
- ZONE (2): arrangement zone; under the effect of the vibrations resulting from the effective energy coming from the zone (1), the grains arrange to a denser state.

6.3 Case of Saturated Soils

Soil response to dynamic compaction in saturated soils is different and of course more complex. Its applicability in saturated layers is generally considered less effective due to the fact that some of the applied energy is absorbed by interstitial water (pore water).

Ménard and Broise [18] introduced the following four factors as the main causes of densification of saturated fine-grained soil layers: high compressibility of water due to the presence of air microbubbles, increase the permeability of the soil mass due to progressive cracks around the impact points, a gradual liquefaction under repeated impacts and a thixotropic recovery. considering the Eq. 4, we deduce the Eq. 7:

$$E' = E.S \tag{7}$$

Moderate levels of stress impulse could cause the collapse of previously loose areas in the dry soil, Saturation means that these cavities are filled with pore water and that the collapse would be inhibited by water resistance to compression. However, the really saturated soils may have few percent of the total void space occupied by air or occluded gas [1, 2]. In this condition, the intensity of the constraints wave would be greatly reduced due to the low bulk of pore water, but the impact stresses still ensure the compacting of the loose structure due to the reduction of the volume of gas in the voids [14, 17]. Given the Eq. 4 we deduce the Eq. 8:

$$Ep = E.(1 - S) \tag{8}$$

As mentioned by Menard and Broise [18], the compressibility of the porous fluid has a significant effect on the behavior of saturated soil under impact load. It is accepted that natural porous fluid contains microbubbles and tiny pockets of air enclosed in voids between soil particles [1, 2]. An accurate assessment of the volume of air pockets trapped in water is difficult microbubbles affect the compressibility of air-water mixtures and thus significantly alter the process of generating porous water pressure. It is therefore essential to take into account the compressibility of the air-water mixture, instead of the compressibility of pure water [1, 19]. Similarly, based on the Eq. 4, determining the energy of the air-water mixture is as follows (Eq. 9):

$$Eaw = Ew.S + Ea.(1 - S) \tag{9}$$

Where; E_{aw} is the energy of the air-water mixture, E_a is the energy of the air, therefore the lost energy is the energy of the air (Eq. 10):

$$Ep = Ea = E.(1 - S) \tag{10}$$

The most determining factor for the ability of a soil type to be improved by dynamic compaction when the water table is near the ground surface is its ability to dissipate excessive pore pressure [1, 15, 19] as illustrates Fig. 16, zone (1). Indeed, when the granular materials extend in the zone (2) Fig. 16, a high proportion of dynamic pulses is transferred to the pore water. If the application of shocks continues before dissipating excessive pore pressures, the pore pressure eventually reaches a level that induces local liquefaction zone (2) Fig. 16.

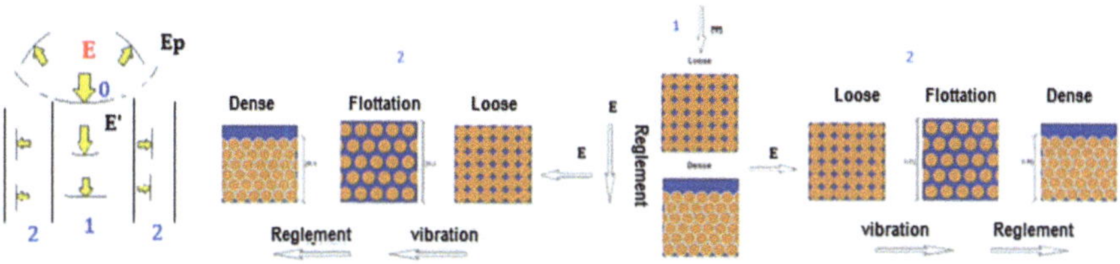

Fig. 16. Overall mechanism of dynamic compaction in saturated granular soils [12].

The overall process of dynamic compaction in saturated granular soils can be separated into three major zones (Fig. 16) [12]:

- ZONE (0): the collision zone; between the tamper and the ground; where the total deformation of the soil matrix, the crushing of the grains during shocks, resulting in a crater and a discharge of soil.
- ZONE (1): the consolidation zone; under the effect of effective energy, the grains densify to a perfectly dense state because of the very fast dissipation of the interstitial water pressure (porous water). The rearrangement goes through two stages: Loose and dense.
- ZONE (2): the liquefaction zone; under the effect of the vibrations resulting from the effective energy coming from the zone (1), grains liquefy and arrange to a denser state. the rearrangement goes through three stages: Loose; float and dense.

6.4 Improved Depth

Menard's popular relationship to predict the depth of improvement is as follows:

$$D = n\sqrt{WH} \tag{11}$$

Where; D is the Depth of Improvement (meter), W is the weight of the tamper (ton), H is the height of fall (meter) and n is an empirical factor that is fundamentally dependent on soil condition.

If we apply our theory in dry/wet soils, given Eq. 11, we get Eq. 12:

$$D = f(E, e) \tag{12}$$

We propose for this equation the following equality:

$$n = e \tag{13}$$

If we apply our theory in saturated soils, given Eq. 11, we get Eq. 14:

$$D = f(E, S) \tag{14}$$

Similarly, for the previous equation, we propose the following equality:

$$n = S \tag{15}$$

When finishes compaction of a point, finding a crater (Fig. 17) whose volume (ΔV) can be defined by Eq. 16:

$$\Delta V = e.\int_0^D V.dD \tag{16}$$

Where; V is the total volume of the compaction point of the soil (meter), S is the surface of the tamper (m^2), Δ H is the height of the crater (mètre) and D is the depth of improvement mentioned in Eq. 11 above.

after development of Eq. 16 and introducing Eq. 13 we deduce Eq. 17:

$$n = \sqrt{\frac{\Delta H}{\sqrt{MH}}} \tag{17}$$

At the end of compacting the whole workshop we found an empirical relationship between the height of the crater and improved depth determined by Eq. 18, where we introduce the proposed coefficient K (Table 1):

$$D = K.\Delta H / K = x.e^{-1} \tag{18}$$

$$x = 6.67.e \tag{19}$$

Table 1. The values of the coefficient x according to the void ratio.

e	e < 0.3	0.3 < e < 0.5	0.5 < e < 0.8	e > 0.8
x	2	3	5	7

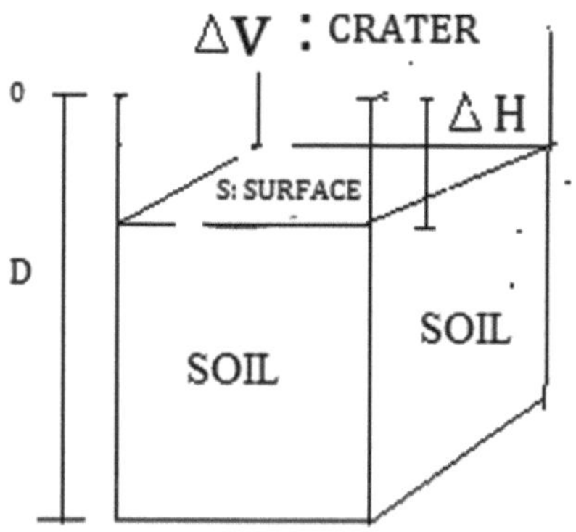

Fig. 17. Phase diagram of a point of compaction of a granular soil by a tamper.

7 Environmental Considerations

Dynamic compaction uses large, highly visible equipment. The process creates noise and vibration, both of which must be considered in Britain under the Law on the Control of Pollution, 1974. By far, the most important consideration is the vibration on the ground. Humans are particularly sensitive to vibration sensing and have a psychological reaction by believing that damage is caused even if the values are well below the established damage thresholds. The prediction of the level of vibration transmitted by the soil is an imprecise science because of the variable nature of the soil characteristics [20]. Careful evaluation is needed when the treated soil is directly in the form of a dense soil that tends to transmit vibrations at greater distances than normal with little attenuation. Pre-existing dense surface or buried layers may have an effect similar to causing transmission of higher vibration levels than expected. The physical performance of the treatment work improves the inter-particle contact of the soils and, as such, the vibration levels can sometimes increase towards the end of the processing operations even though the final energy levels are significantly lower than those made for the initial tamping passes [20].

If we take all the equations from 1 to 10 above we can easily obtain the graphs of Fig. 18; which is clearly illustrated; that the effective energy is proportional to the degree of saturation and the void ratio as opposed to the lost energy. It can be concluded that the soil response to the energy impact depends on the void ratio and degree of saturation. Thus, the level of compaction efficiency depends on the effective energy, too, the lost energy is minimal in saturated soils.

In dry/wet granular soils, the graph of the energy response is divided into four parts as shown in Fig. 19a to give us an approach to estimate the effectiveness of dynamic compaction and its effect (vibrations). In saturated soils; Practically the amount of air in the porous water generally varies at most between 15 to 25%; which implies a degree of saturation between 75 and at least 85% (Fig. 19b), which limits the applicable field in saturated soils, and we can consider the rest as a field not practically applicable. The most important remark is that the lost energy in the applicable field is negligible.

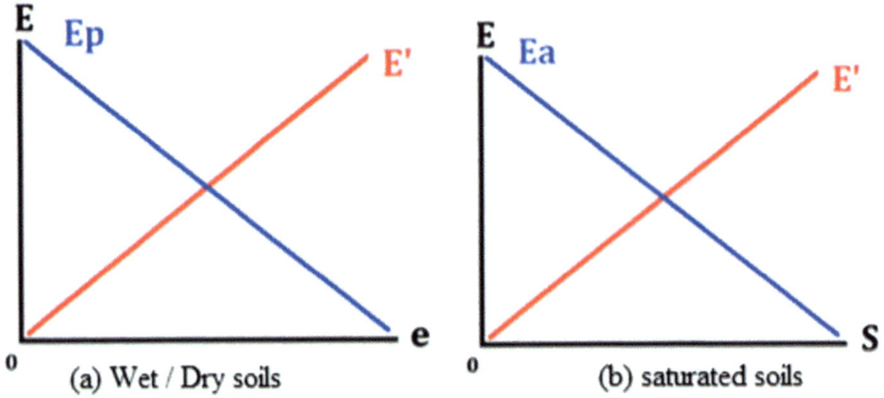

Fig. 18. Graphs of the energetic response according to the state of the granular soil according to our theory.

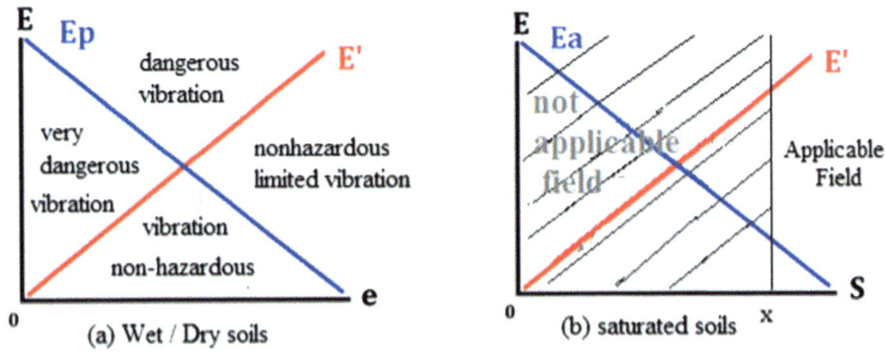

Fig. 19. Fields of application and effects of DC according to the graphs of the energy response of our theory.

8 Discussion of Results

The optimal number of blows in the second test board is greater than the first, because of the embankment height of the first PE-1 test board which is very small compared to the second PE-2, which implies that the void ratio in the first board PE-1 column is lower (at least) to the second board PE-2. In addition, the water level in the second board PE-2 is high, against it does not exist in the first board PE-1. Therefore the lost energy (Ep) in the first PE-1 zone is higher compared to the second one (Eqs. 3 and 6), which explains the appearance of cracks in the harbor administration building. In areas where the water level is high, there is no significant vibration in the vicinity of the column to be compacted, since the majority of the effective energy (E′) absorbed by the porous water (Eq. 4) and the lost energy (Ep = Ea) are negligible, which explains the stability of the blocks during the compaction of the penultimate line (la Figs. 5 and 11).

9 Conclusion

Dynamic compaction is a method of soil improvement that has been widely used because of its cost-effectiveness, simplicity and the considerable depth it affects, however, very few theoretical models are available, which prevents us from understanding all the side effects resulting from it.

In this paper, a theory has been proposed concerning the response of the soil to a high energy impact during dynamic compaction. This theory gives us a broad understanding and precision during the improvement, which gives us a wider field, for interpreting the results associated with dynamic compaction. We also observe that the equations obtained through our theory are considered acceptable and make it possible to facilitate the calculations of certain parameters, which gives a great credibility to our theory.

When the lost energy (Ep) complicates the problem, there are three main methods to reduce its effect, the first is to simply reduce the drop height and to compensate by increasing the number of strokes per footprint, this reduces both the impact energy and penetration of the stress pulse that may have reached a dense layer underlying. The second reduction method is to use a smaller weight and the third to dig a cut trench to a depth sufficient to intercept the surface wave. When we cannot implement these solutions for technical reasons, we should then create something that can absorb the lost energy (Ep) without affecting or decreasing the efficiency of the effective energy (E′).

References

1. Lehuérou-Kérisel, J., Caquot, A., Kersiel, J.: Traité de la mécanique des sols, 4eme édn. Paris (1966)
2. Tarzaghi, K., Peck, R.B., Mesri, G.: Soil Mechanics in Engineering Practice, 3rd edn. Wiley, Hoboken (1996)
3. Andrus, R.D., Chung, R.M.: Ground improvement techniques for liquefaction remediation near existing lifelines. Report, NISTIR 5714, Building and Fire Research Laboratory, National Institute of Standards and Technology, Gaithersburg (1995)
4. Van‘t Hoff, J., Van Der Kolff, A.N.: Hydraulic Fill Manual for Dredging and Reclamation Works. CRC Press, Taylor & Francis Group, Boca Raton (2012)
5. USACE: Engineering and design: dredging and dredged material management. U.S. Army Corps of Engineers, Manual No. 1110-2-5025 (2015)
6. Bell, F.G., Detry, T.V.: Méthode de traitement des sols instables. Eyrolles, Paris (1978)
7. Moseley, M.P., Kirsch, K.: Ground Improvement. CRC Press, Taylor & Francis Group, Boca Raton (2004)
8. Hamidi, B., Nikraz, H., Varaksin, S.: A review on impact oriented ground improvement techniques. Aust. Geomech. **44**(2), 17–24 (2009)
9. Khelalfa, H.: Coastal Soil Treatment to Stabilize Vertical Breakwaters. LAMBERT Academic Publishing (2018). ISBN: 978-3-330-03812-7
10. Lukas, R.G.: Dynamic compaction for highway construction, vol. 1, design and construction guidelines. Report No. FHWA/RD-86/133, Federal Highway Administration, Washington, DC (1986)

11. Ebid, A.M.: Simple mathematical approach to simulate granular fill behavior under dynamic compaction. Int. J. Eng. Innov. Tech. **5**(5), 24–31 (2015)
12. Khelalfa, H.: Dynamic compaction with hight energy of sandy hydraulic fills. Sci. Rev. "JMES" **4**(3), 121–138 (2017)
13. Vernay, M., Morvan, M., Breul, P.: Etude du comportement des sols non saturés à la liquéfaction. 33èmes Rencontres Universitaires de Génie Civil, Bayonne (2015)
14. Jefferies, M., Been, K.: Soil Liquefaction, a Critical State Approach, 2ème edn. CRC Press, Taylor & Francis Group, Boca Raton (2015)
15. Ghassemi, A., Pak, A., Shahir, H.: A numerical tool for design of dynamic compaction treatment in dry and moist sands. Iran. J. Sci. Technol. Trans. Civil Eng. **33**(4), 313–326 (2009). https://doi.org/10.22099/ijstc.2009.709
16. D'Appolonia, E.: Loose sands – their compaction by vibroflotation. In: Symposium on Dynamic Testing of Soils, 56th Annual Meeting of the American Society for Testing Materials, ASTM Special Technical Publication No. 156, Philadelphia, PA, pp. 138–162 (1953)
17. Jafarzadeh, F.: Dynamic compaction method in physical model tests. Sci. Iran **13**(2), 187–192 (2006)
18. Ménard, L., Broise, Y.: Theoretical and practical aspects of dynamic consolidation. Geotech. **25**(1), 3–18 (1975). https://doi.org/10.1680/geot.1975.25.1.3
19. Ghassemi, A., Pak, A., Shahir, H.: Numerical study of the coupled hydro-mechanical effects in dynamic compaction of saturated granular soils. Comput. Geotech. **37**(1–2), 10–24 (2010). https://doi.org/10.1016/j.compgeo.2009.06.009
20. Moseley, M.P., Kirsch, K.: Ground Improvement. Spon Press, New York (2005). ISBN 0-203-57085-5

Contaminated Soil Remediation: A Novel Cartographic Model for the Representation of Technical Maps

Salvatore Barba[1(✉)], Vincenzo Naddeo[2], Paula Mardisa[3], and Vincenzo Belgiorno[2]

[1] Department of Civil Engineering, University of Salerno, Fisciano, Italy
sbarba@unisa.it
[2] Sanitary Environmental Engineering Division, University of Salerno, Fisciano, Italy
{vnaddeo, v.belgiorno}@unisa.it
[3] School of Architecture, National University of Córdoba, Córdoba, Argentina
paulamardisa@gmail.com

Abstract. In the projects of remediation of contaminated site is necessary to produce several technical and cartographic maps that must describe the plan of characterization of the area, the level of the contamination, the sources and the possible techniques for its remediation. Generally, each project uses different ways for the representation of the variables causing an extraordinary effort by technicians of the Institutions that need to use these projects. In this context, one important challenge is the definition of a unifying language in order to communicate the main technical and specific indications in an effective way.

The aim of this work is the proposition of a graphic model useful for the cartographic representation of all documents needed in the field of contaminated sites remediation. According a standardized graphic code was created which signs and symbols that represent the different variables detected in the analysed issue compose. In order to be able to define this model, it was fundamental to reach a great level of synthesis, identifying the relevant information necessary to communicate technical information in a graphical way, trying to produce an effective and clear univocal message.

The proposition was implemented in a case study localized in South Italy. This implementation will be used for the improvement of the developed graphic model and to show its functionality.

1 Introduction

Communication and representation have today a main role in our society, this is the base behind the idea to present a proposal for a graphic model associated with remediation activities of contaminated soils.

Today, information are available via many channels and in various different types; how the user reads and understands those information defines if the message was correctly sent or not.

H. Ameen and T. Sorour (Eds.): GeoMEast 2018, SUCI, pp. 105–114, 2019.
https://doi.org/10.1007/978-3-030-01929-7_8

Therefore, a new graphic model aiming to improve the efficacy of the communication, based on the potentiality of all the properties of the language, is proposed.

The visual language is a graphic channel of communication that sends information to a specific user by symbols and signs. The sent information is a synthesis and simplification of some specific phenomenon.

Based on this reduction, a graphic model, therefore, will never transmit the entire information.

The result will be a cartography map of contaminated soils, containing a group of codified symbols.

These symbols will use as many properties of any element as possible, and, by considering graphic variables, they will determinate the best way to communicate the represented information, without ambiguities.

A case study will solve the deep analysis of the specific activity, needed to understand which kind of information should be transmitted, defining in this way the graphic model.

The first step will be to study the theory of the subject-problem to represent. By defining what should be transmitted by the cartography map, it will identify the necessary variables to be selected.

It is an essential key point for this work to identify which information can be represented. As each symbol will have a one-one correspondence with each element, (defining for example a category, a type, a variation), the more precise and brief is the data, the higher efficient is the communication.

To understand better how to create this graphic model, the work is divided and structured in two parts: WHAT (defining the activity of characterizing a remediation project for contaminated soils) and HOW (defining the graphic models and the study of communication).

2 Composition of a Technical-Graphic Message

The composition of the message will be the right organization of agreed symbols (that will have a specific meaning), by using it together or isolated.

A special consideration is the possibility to join different symbols, but, being careful that the most convenient images are the most geometrically ones.

The graphic variables are the key to the interpretation of the same symbol but in different contexts, which means that depending on the position and disposition, the same symbol may assume different meanings.

We could summarise this section, saying that, the less quantity of variables used, the higher essential index, will be the graphic message efficient and readable.

3 Methodology

Firstly, the variables are defined starting from the collected data during the characterizing activities of the case study "Piazza Castello, Avellino, Italy". Secondly, these variables are organized by layers to improve their readability, defining macro-groups, where interconnected elements share common characteristics. Thirdly, after listing variables, we will define signs and symbols.

The entire graphic model has the same language, with uniformity, expressing a shared position in front of the problematic, where each individual symbol belongs to that cartography, to the specific case study. We will describe this in more details, in the next section.

4 Graphic Model: Generalities

Thinking with the uniformity of the used language, there are some parameters from where symbols were designed. Once these parameters are defined, the game criteria to apply them, is the differentiation element for each specific use, for example:

Predominantly, symbols are composed by lines, dots and surfaces, looking for a simple geometry, trying to not confuse the reader.

Symbols were designed like "icons" or "motivated type", trying to imitate the same geometrical shape than reality, giving the possibility to provide a quick lecture without the need to read any legends. For non-motivated symbols, existing conventions were used. The legends aim to be used in very specific cases.

For colour variable, the criteria are based via the usage of tones, transparency and saturation, always looking for the simplest expression for a clean lecture (Fig. 1).

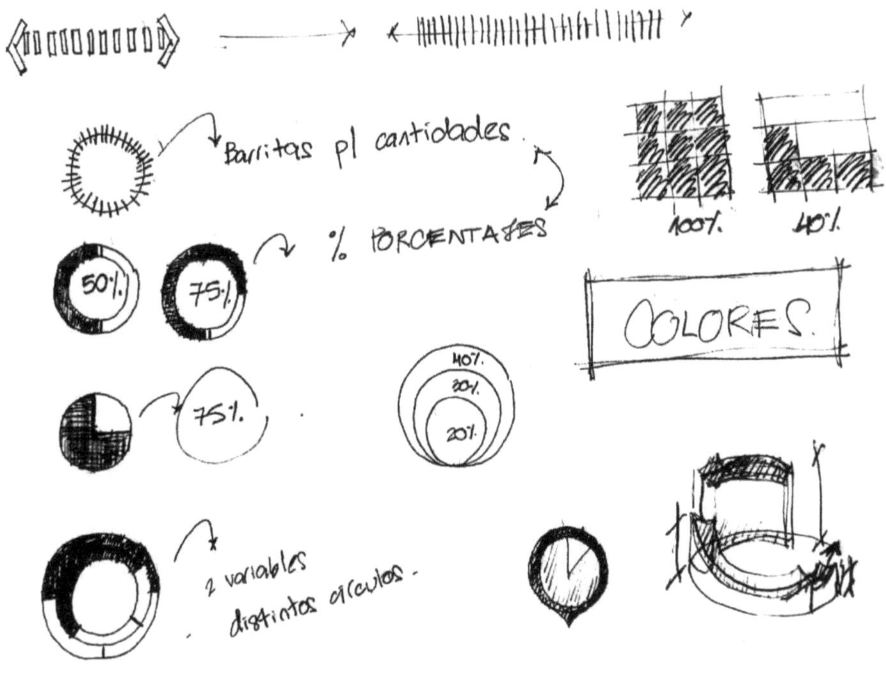

Fig. 1. Design process - handmade analogical researches

Two main categories were introduced: static and dynamic symbols. The first one, includes symbols that the user can apply directly to the cartography (general use), while the second one, includes symbols that, applying some logical rule, can be built in the moment according the specific case (specific use).

A floor plan or a section can define the same information, so different symbols for each technical draw can do the same.

Generically, a symbol works for a specific scale, not easily adapting to other scales. For Piazza Castello's case study, the scale is same than the original project: from 1:350 to 1:1000.

The standard graphic code is one of the goals of this work. In order to achieve it, the proposal should be applicable to more than one case study. Thinking in this way, the model could be extended to a generality. Looking then a unified language, its readability will be improved, giving the right communication in any case.

5 Graphic Model: The Creation of Symbols

Starting from the idea to create a unified language, it is important to highlight the used software:

- *Adobe Illustrator*, being a vector-based program, is one of the better options that fits to draw symbols and mix them to create the cartography. In this case was used for the entire cartography of Piazza Castello.
- *ColorBrewer 2.0*, a web-based application that standardizes and extracts the card of in-cartography used colours (Fig. 2).

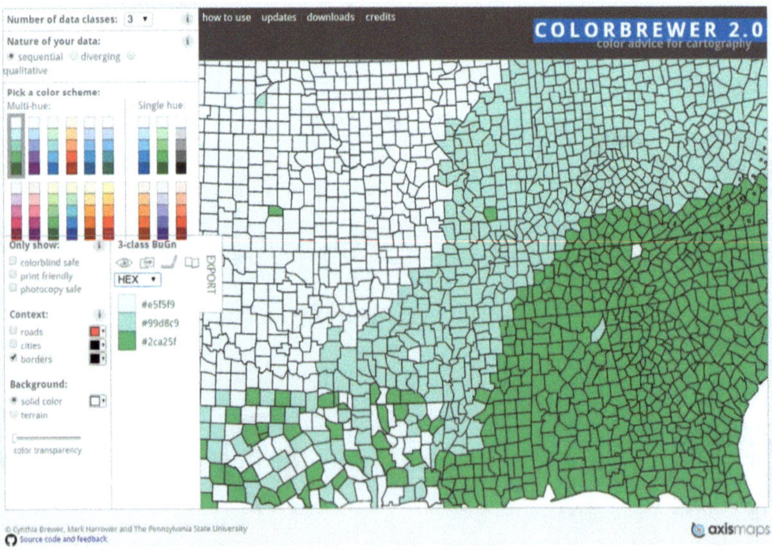

Fig. 2. Main window of ColorBrewer 2.0 application

The main goal of ColorBrewer (http://colorbrewer2.org) – without the need of great webGL resources – is to give a group of related colours for cartography, selecting them by their readability either alone or mixed. The result guarantees a very clear reading even if the drawing scale is relevantly small. Some options of this application are:

- *number of data classes*, a class is a category in which our variables are divided. The higher is the number of classes, the higher is the quantity of contained information, but at the same time too many classes could generate confusions to the reader;
- *nature of data*, the application will generate a different colour scale according to the kind of selected data that could be sequential, divergent or qualitative;
- *pick a colour scheme*, from a very high range;
- *pick a colour class*, could be CMYK, RGB and HEX values. The destination of the cartography (digital or printed) could define the best option. The offered mixes are compatible between them and from one system to another.

A software like this reveals a very high importance in the role to defining something universal, because it is working with standard colours and conventions making the knowledge sharing much easier.

6 Towards a Proposal of a Graphic Model

After processing and compering, the collected information from the Piazza Castello's case study (texts and old cartographies) with the presented variables, an analysis of the used graphic language for the characterizing is presented.

From this comparison, it was deduced that written information is not always capable to translate the graphical information, or the presented order in the cartography. Therefore, one of the objectives of this work is to create a language that allows the complete representation of all information, introducing the idea to use layers, connecting data from different categories.

In the graphical language, the main detected problem is the consistency and coherence between varieties of cartographies. Confusion is produced using different colours when trying to represent the same variable, negatively affecting the efficiency of the communication.

The case study did verify that the previous presented need to interpret the symbol with the legend, caused for an incomplete representation of the data within the symbol itself.

A brief list of the main identified problematics for Piazza Castello's example are:

- missing graphical information and with no complete order;
- missing coherence and unity in language between different cartographies;
- mandatory use of legends required to understand in full what a symbol is trying to represent.

Based on those problematics, the new graphical model was proposed.

7 Application of the Model to Cartographies

Following the founding of the problematic illustrated and its specific issues, the new developed symbols and variable (ready to be drawn an organized), are used to produce and code the cartography from Piazza Castello's example.

Next step will be to illustrate tables with the proposed layers; these layers organize the cartographies with their variables.

It is important to emphasize that the layers' variables can be overlapped or broken down, based on the desired outcome and purpose. Each layer will represent one thematic, so different meanings could be obtained by overlapping them.

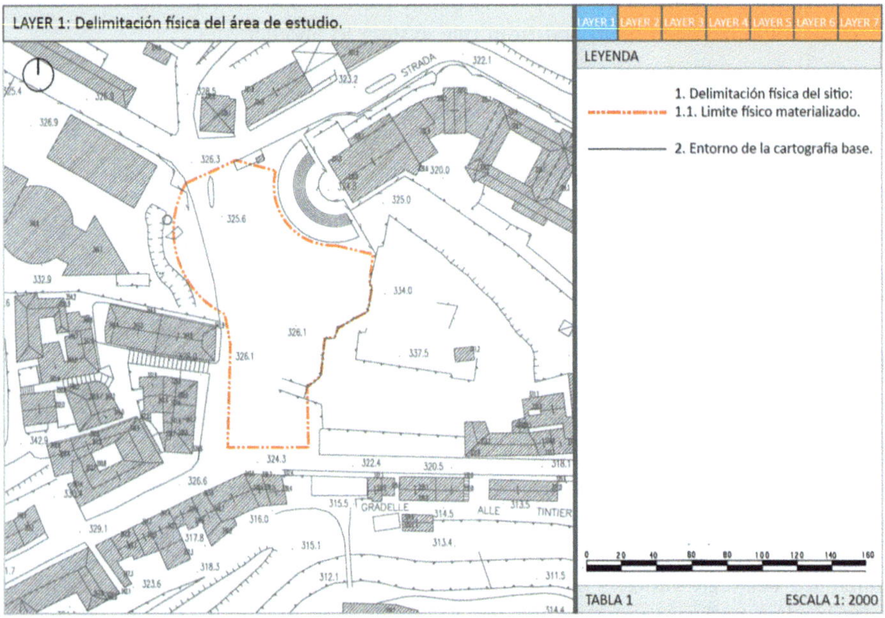

Fig. 3. Layer 1 - case study area's limits

Variables are categorised and divided in the following:

- Layer 1, delimitation of the area (Fig. 3);
- Layer 2, historical cartography (Fig. 4);
- Layer 3, physical-spatial elements of the site (Fig. 5);
- Layer 4, field's components by particle size distribution and nature (Figs. 6 and 7);
- Layer 5, sampling activity (exploration) (Fig. 8);
- Layer 6, sampling activity (section);
- Layer 7, extension of the contamination (Fig. 9).

The complete model presents a group of symbols that compose the code: each symbol is organized in two specific tables: the first one containing their graphical

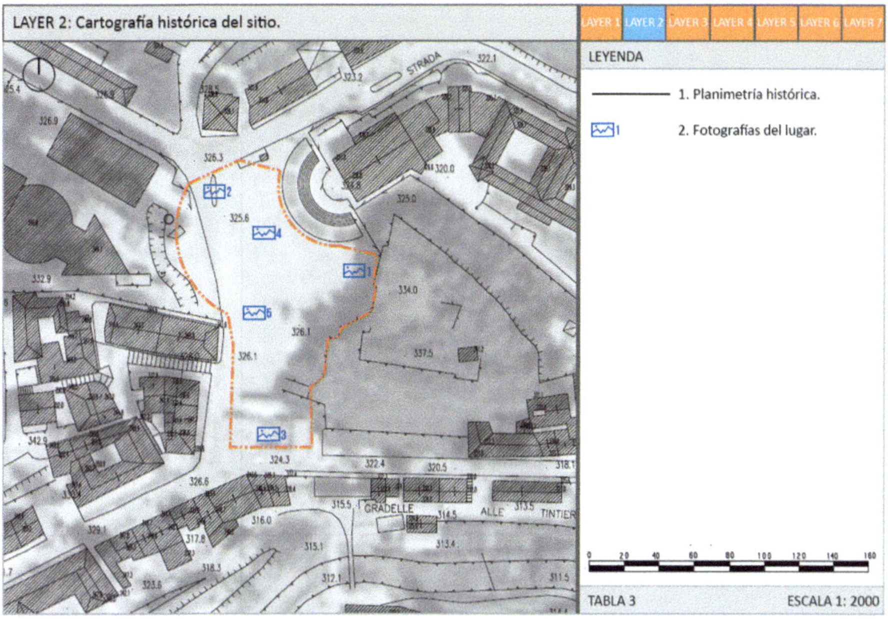

Fig. 4. Layer 2 - historical cartography

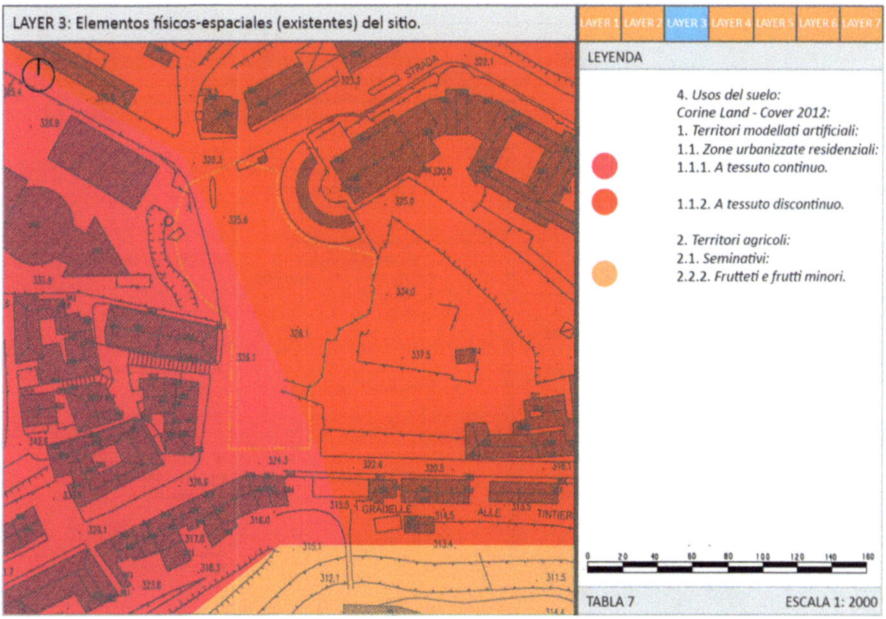

Fig. 5. Layer 3 - existing physical-spatial elements of the site

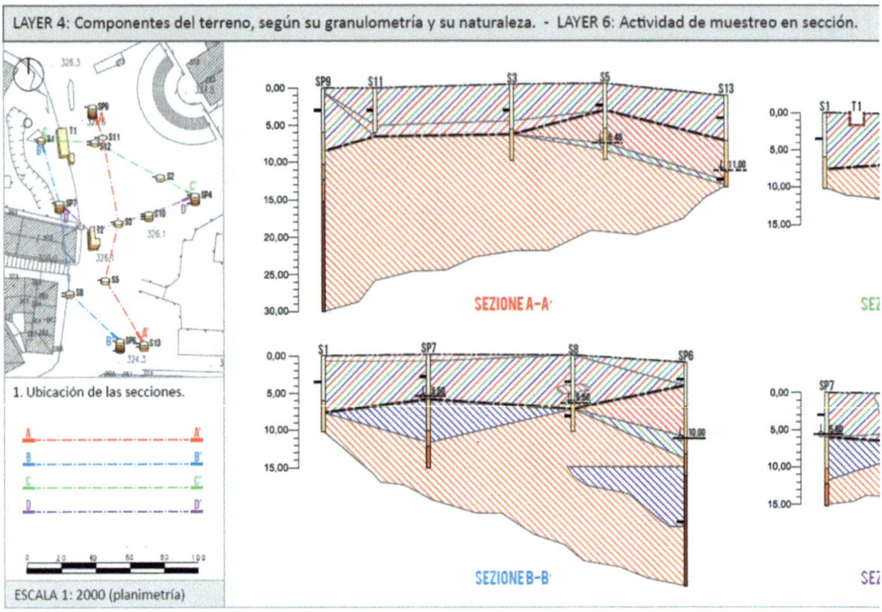

Fig. 6. Layer 4 & 6 - sampling activity (section)

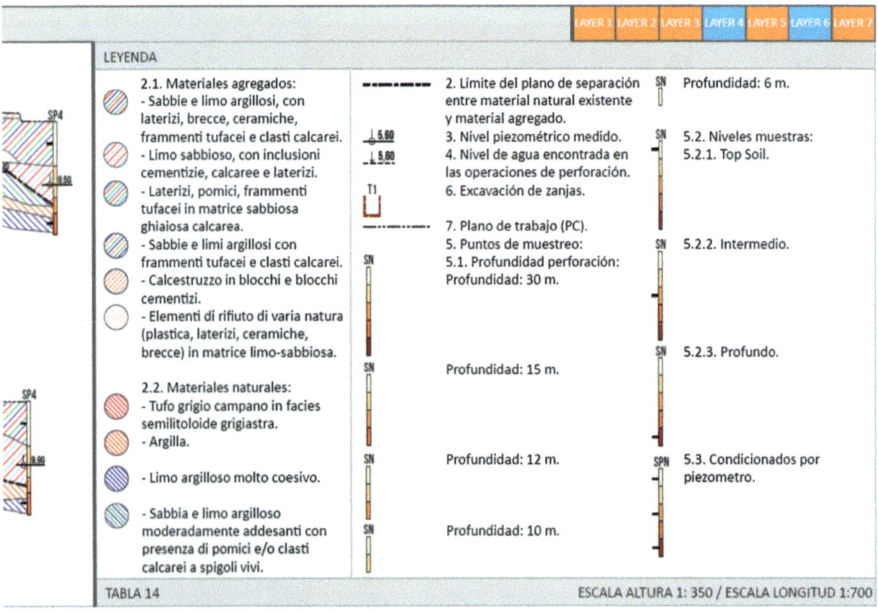

Fig. 7. Layer 4 & 6 - ground's components by particle size distribution and nature

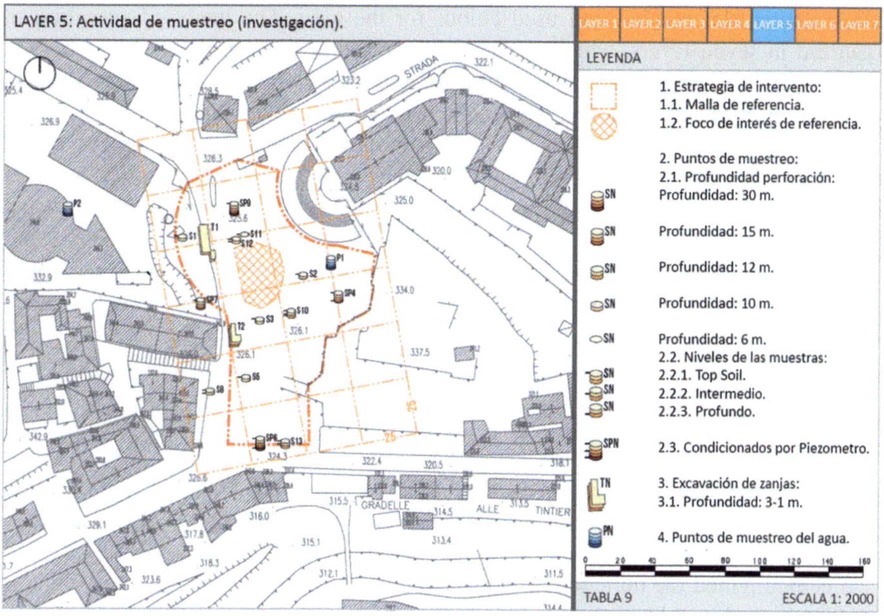

Fig. 8. Layer 5 - sampling activity (research)

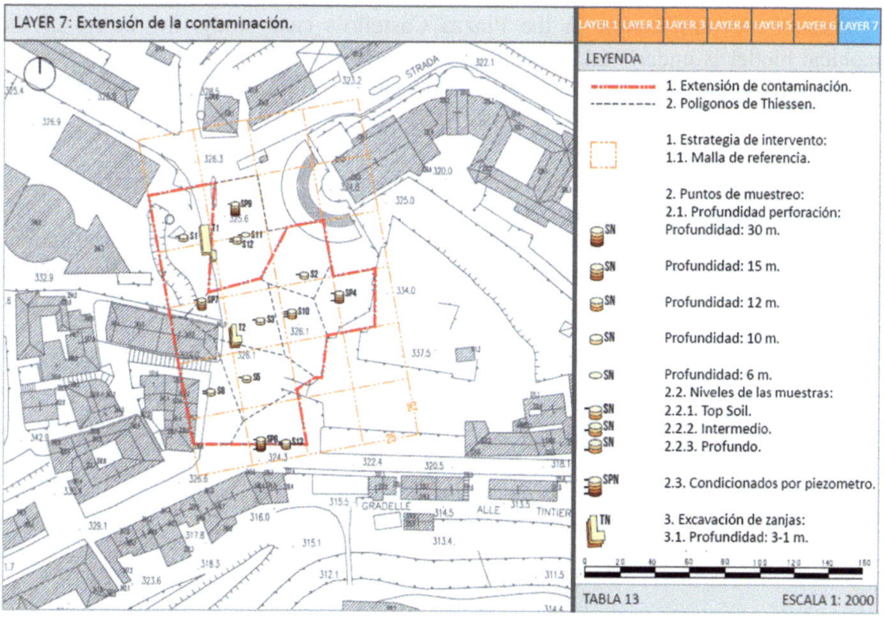

Fig. 9. Layer 7 - contamination's extension

representation, designation and used colour for the case. The analysed activities are organized in seven layers, which are identified at the beginning of the research.

The second table describes the composition of the symbol, its elementary unit, the implantation in the cartography, the type and its graphical variables.

Reminding the difference between static and dynamic symbols, a template can be defined for each dynamic symbol, while just one is necessary to all static ones. The need of flexibility within dynamics symbols is crucial for the adaptation of this system to different cases.

It is important to understand how tables are built, to give the model the possibility to add and modify new variables once detected, keeping the same logic and language within the proposal.

8 Conclusion

The presented model can successfully be used for the representation of the characterizing activities for contaminated soils, giving a coherent, efficient and faster-comprehended model during reading of cartographies.

It is highlighted the no-legend characteristic need for the interpretation of symbols. Even if a legend could be important to clarify information or to specify some variables and their content, it is not compulsory for a first general approach of the transmitted message. The finality of the cartography is to represent a specific variable, while the legend may give us further details.

Through the application in the Piazza Castello's case study, the utility of the graphical model is underlined, presenting a uniform language in all the cartographies, which shows, coherently and efficiently, information in one unique axis, which could be deemed universal.

Acknowledgments. Funding Information: Application of Advanced Membrane Desalination for a sustainable development of water resources in Mediterranean Countries (Projects for Exchange of Researchers selected within the frame of the Executive Programme of Scientific and Technological Cooperation between the Italian Republic and the Arab Republic of Egypt for the years 2016–2018).

Chemical Transport Through Porous Media for EOR of Upper Assam Basin, India

Joyshree Barman[1] and Subrata Borgohain Gogoi[2(✉)]

[1] Department of Petroleum Technology, Dibrugarh University, Dibrugarh 786004, Assam, India
joyshreebarman2012@gmail.com

[2] Petroleum Technology, Dibrugarh University, Dibrugarh 786004, Assam, India
subrata@dibru.ac.in

Abstract. In Alkaline Surfactant Enhanced Oil Recovery (ASEOR) an alkali and surfactant/surfactants are used to recover the residual oil that remains after secondary brine flooding. The alkali, which is Sodium Hydroxide (NaOH) in this case, reacts with acidic components in the crude oil to form surface-active substances. A GC-MS spectrum of Upper Assam crude oil reveals the presence of carboxylic acid groups leading to in-situ formation of surfactants, which in turn decreases the interfacial tension (IFT) between the oleic and aqueous phases for better oil recovery. While the anionic surfactants used were Black Liquor (BL) and Sodium Dodecyl Sulphate (SDS). The Critical Micellar concentration (CMC) of 0.6 v/v % BL, 0.4 v/v % of 0.1 M SDS and mixed 0.6 v/v % BL + 0.4 v/v % of 0.1 M SDS one at a time was added NaOH to enhance the effectiveness of NaOH in further decreasing the IFT of the Alkali-Surfactant (AS) slugs. The core flooding experiment were conducted by slugs 0.6 v/v % BL + 0.6% NaOH, 0.4 v/v % of 0.1 M SDS + 0.6% NaOH and mixed slug 0.6 v/v % BL + 0.4 v/v % of 0.1 M SDS +0.6% NaOH. The recovery of residual oil after secondary flood was found to be more in case of mixed slug 0.6 v/v % BL + 0.4 v/v % of 0.1 M SDS +0.6% NaOH about to 32.04% (PV 4.87 cc) out of IOIP (PV 15.2 cc). The paper also make an attempt to study the both dynamic and static adsorptive nature of the AS slugs and found that dynamic adsorption (mg/g) was much less compared to static adsorption (mg/g). And also addition of NaOH to each formulated slug it enhanced the adsorptive nature on PM in terms of economic. The four isotherms: Langmuir, Freundlich, Temkin and Linear were investigated for the best recovery mixed slug 0.6 v/v % BL + 0.4 v/v % of 0.1 M SDS +0.6% NaOH. The Freundlich isotherm found to be best fit adsorption isotherm based on regression co-efficient (R^2).

Keywords: Surfactant · Alkali · ASEOR · GC-MS · IFT Adsorption · Isotherms

Nomenclature

ASEOR	Alkaline Surfactant Enhanced Oil Recovery
2θ	Bragg angle
BL	Black Liquor
c_0	Initial concentrations of adsorbate, mg/L

H. Ameen and T. Sorour (Eds.): GeoMEast 2018, SUCI, pp. 115–136, 2019.
https://doi.org/10.1007/978-3-030-01929-7_9

c_e	Equilibrium aqueous concentration of surfactants, mg/L
CMC	Critical Micellar concentration
EOR	Enhanced oil recovery
IFT	Interfacial tension
K_H	Linear constant, L/m^2
m	Weight of the adsorbent, g
NaOH	Sodium Hydroxide
N_c	Capillary number
q_e	Amount of adsorbate adsorbed on the adsorbent, mg/g
q_0	Maximum monolayer coverage capacity, mg/g
SDS	Sodium Dodecyl Sulphate
B	Temkin constant
PM	Porous media
SDS	Sodium dodecyl Sulphate
Ca^{2+}	Calcium ion
ppm	Parts per million
w/v %	Weight/volume %
H_3O^+	Hydronium ion
XRD	X-ray diffusion
KOH	Potassium Hydroxide
NaCl	Sodium Chloride
n	Sorption intensity
K_L	Langmuir isotherm constant, L/mg
$NaHCO_3$	Sodium bicarbonate
rpm	Rotation per minute
Mg^{2+}	Magnesium ion
K_f	Freundlich adsorption constant, mg/g
GC-MS	Gas Chromatography – Mass spectrometry
K_t	Equilibrium binding

1 Introduction

Currently, most oil fields in Assam are in the later production period and the water cut increases rapidly, even to more than 90% especially in the fields of Moran and Nahorkatiya (Gogoi 2009). Waterflooding no longer meets the demands of oilfield production. Thus, it is inevitable that a new technology will replace waterflooding. The new technique of ASEOR flooding has been developed on the basis of earlier chemical EOR research pertaining to Upper Assam Basin (Gogoi 2010). ASEOR flooding uses the benefits of two flooding methods simultaneously, and oil recovery is greatly enhanced by decreasing IFT and enhancing microscopic displacing efficiency (Gogoi 2009; Shen and Yu 2002; Wang et al. 2000; Wang et al. 2002; Sui et al. 2000).

Recently, much intensive research has been done on ASEOR flooding both in India and worldwide, achieving some important accomplishments that lay a solid foundation for the extension of this technique to practical application in oil fields. The acid number and the acid group of crude oil were determined to examine the applicability of alkali as an EOR chemical. Various IFT studies were done to determine the best surfactant. The flooding experiments were conducted to examine the recovery efficiency of crude oil from the underground porous media. A detailed adsorption studies were conducted to find the adsorption of chemicals on the porous media. For a vertical heterogeneous reservoir, ASEOR flooding increases displacement efficiency by displacing residual oil through decreased IFT, simultaneously improving sweep efficiency by extending the swept area in both vertical and horizontal directions. Some physical and chemical phenomena, such as emulsion, scale deposition, and chromatographic separation, occur during ASP flooding (Arihara et al. 1999; Guo 1999), because ASEOR flooding in porous media involves many complicated physicochemical properties, many oil-recovery mechanisms still need to be investigated.

2 Materials and Methods

2.1 Materials Used

Tables 1 and 2 showed materials and instrument used during experiment work.

Table 1. Showing materials used during the experiments

SN	Sample	Specifications	Source
1	BL, main constituent is Na-lignosulfonate	Anionic surfactant, M.W. = 368 gm/mol	Nagaon Paper Mill, Jagiroad, Assam
2	SDS ($C_{12}H_{25}OSO_2ONa$)	Anionic surfactant, M.W. = 288.37 gm/mol	Merck Life Science Private Limited, Mumbai
3	NaOH	M.W. = 40.0 gm/mol	RFCL Ltd., New Delhi
4	Crude oil	Sp. Gr. 0.916, 23^0 API, acid no. 1.76 mg KOH/gm	Upper Assam Basin
5	Porous media (PM)	Conventional core samples Depth = 3536 m	Oil E & P, Assam
6	Acetone, C_3H_6O	M.W. = 58.08 gm/mol	SDFCL, Mumbai
7	Ethanol, C_2H_5OH	M.W. = 46.07 gm/mol	Biochem Life Sciences, New Delhi
8	Ethyl Acetate, $C_2H_5COOCH_3$	M.W. = 88.1 gm/mol	
9	Liquid Paraffin Light,	Wt. per ml at 20 °C is 0.83–0.86	Fisher Scientific, Mumbai
10	Sodium chloride, NaCl	M.W. = 58.44 g/mol	Merck Life Science Private Limited, Mumbai

Table 2. Showing the instruments used during experimental works

SN	Instrument name	Make	Model	Purpose
1	Spinning Drop Tensiometer	USA	Grace M6500	IFT
2	Rotospin	TARSONS	CAT: 309OX	Well mixture
3	UV-Spectrophotometer	Schmadzu	UV-1800	Static and Dynamic adsorption
4	Sieve shaking machine	S.D. HARDSON & Co		mesh size
5	XRD	Ridaku	Ultima IV	Mineralogical Composition
6	Magnetic Stirrer	Tarsons	External probe Pt-1000	Mixture
7	Centrifuge	EURO Diagnostic Systems Pvt. Ltd	ELECTRA LAB CENTRIFUGE	0-8,000 rpm
8	GC-MS	Agilent	Agilent GC 7820 A & Agilent MS 5920 MS	Organics
9	Dehumidifier control oven	Vindish Instruments Pvt. Ltd.	101	Humidity
10	Electronic Weight Balance	Mettler Toledo, Switzerland	A122	Weight
11	Core flooding system	Gon Engineering works Dhanbad, Jharkkand		Dynamic adsorption & recovery of residual oil

2.2 Methods

Acid no. of Crude Oil

The standard method ASTM D8045-17 was used to determine the acid number of crude oil. 0.1 N of KOH solution was taken in a burette. Titration experiment was done by pouring KOH in a drop wise manner in 20 ml solution of ethanol (C_2H_5OH) and 2 drops of phenolphthalein indicator in a conical flask, until the solution becomes stable pink color indicating the end point of titration i.e., it attained the neutral point. Then 1 gm of the crude oil is added to the neutral solution of the conical flask and again titrated with 0.1 N of KOH from burette until wine red color appears. The mathematical formula for acid number is as follows:

$$\text{Acid number} = \frac{vol.\,of\,burette\,reading \times Normality\,of\,KOH \times Mol.\,wt.\,of\,KOH}{wt.\,of\,oil}$$

Acid Group Present (Alkyl Group) in Crude Oil Determined by GC-MS
The organics in crude oil as in Table 1 SN 4 were determined by GC-MS. The samples were prepared by dissolving in solvent ethyl acetate ($C_2H_5COOCH_3$) until the solution was colorless. The solution is then injected to the GC and GC peaks obtained were analyzed by NIST library data from Agilent GC-7820A and Agilent MS 5920 (Agilent GC-7820A and Agilent MS 5920 2007).

Preparation of Surfactant Solution for IFT Measurement
IFT experiments were done in Spinning Drop Tensiometer to determine the dynamic IFT between the oleic and the aqueous phases. The aqueous samples were prepared in 50 ml tarsons in concentrations from 0.1% to 1% v/v. The stock and the experimental aqueous phase samples were prepared as shown in Tables 3, 4 and 5. 0.1 M SDS and BL were added to that 3000 ppm brine in volumetric tarson tubes. The samples were well in Rotospin at 50 rpm for 48 h.

Table 3. Stock aqueous phase sample preparation for SDS

SN	Sample	Amount of SDS, gm	DW, ml
1	0.1 M SDS	1.441	50 ml

Table 4. Experimental aqueous phase samples from stock aqueous sample

SN	Conc. of Surfactant (BL/0.1 M SDS), %	Vol. of Surfactant (BL/0. 1 M SDS), ml	Vol. of 3000 ppm brine, Ml
1	0.1	0.05	49.95
2	0.2	0.1	49.90
3	0.3	0.15	49.85
4	0.4	0.2	49.80
5	0.5	0.25	49.75
6	0.6	0.3	49.70
7	0.7	0.35	49.65
8	0.8	0.4	49.60
9	0.9	0.45	49.55
10	1	0.5	49.50

Table 5. Experimental samples of NaOH conc. (W/V %) with CMC of BL/0.1 M SDS

SN	Conc. of Alkali (NaOH), %	Wt. of Alkali NaOH), gm	Vol. of CMC concentration (BL/0.1 M SDS), ml
1	0.1	0.05	50
2	0.2	0.1	50
3	0.3	0.15	50
4	0.4	0.2	50
5	0.5	0.25	50
6	0.6	0.3	50

After experimentally determining the IFTs, the CMCs of the surfactants were determined from the plots of IFTs and surfactant concentrations. After getting CMC value of SDS and BL in 3000 ppm for the samples as shown in Table 4, different concentration of NaOH were added to that CMC of SDS and BL as in Table 5. IFTs were measured at 75 °C under atmospheric pressure for different formulations of aqueous phases containing SDS + NaOH, BL + NaOH and SDS + BL + NaOH with a single oleic phase which in this case was acidic crude as determined by GC-MS.

The equilibrium IFT was calculated using from the equation as available in the Spinning Drop Tensiometer Grace M6500.

$$\gamma = 1.44 \times 10^{-7} \times \Delta\rho \times D^3 \times \theta^2 \quad (1)$$

Where

γ = IFT, mN/m
$\Delta\rho$ = Density difference between oleic and aqueous phase, gm/cc
D = Diameter of the oil bulb, mm
θ = Rotation, rpm

Core Sample Preparation for Coreflood, XRD and Adsorption Study
The core was cut at core cutting machine and cleaned by Soxhlet extraction using 1:1 toluene and methanol, and then dried at 100 °C until constant weight attained. The core dimensions, weight, porosity and permeability were measured.

Porous media as in Table 1 SN 5 was sieved in the sieve shaking machine to obtain 60–70 mesh size sand grains for adsorption study and the powder collected in the PAN was taken for XRD analysis. The 60–70 mesh size sand grains were washed with distilled water (DW) for five times by the process of settling and decanting. After removing the suspended particles, the residual wet sand grains were dried at 80 °C in the dehumidifier till a constant weight was obtained. The adsorbents which were the 60–70 mesh size sand grains were used to determine the SDS, BL and NaOH adsorption by UV-Spectrophotometer.

After powder collected in the PAN were taken for pipetting method. In pipetting method, PAN collected fine sand powder were mixed with 1000 ml DW in 1 litre measuring cylinder and mixed properly by a magnetic stirrer for 5–10 mints and kept undisturbed for an hour. The supernant, about 5 ml of the solution was collected in small beaker. The supernant collected from the solution was from the center of the level of solution from top, since clays are lighter than other minerals (Hurst 1999). This process was continued until considerable amount of sample was collected that was sufficient for XRD experiment. XRD of the prepared sample was recorded in a range of Bragg angle ($10^0 \leq 2\theta \leq 75^o$) using Ridaku XRD measuring instrument and analysed (Tucker 1998).

Flooding Experiment
The core was initially saturated with brine and brine was flooded through the core to measure the absolute permeability (K_a) of the core by steady state Darcy's law. It was then flooded with oil to measure the effective permeability of oil (K_{eo}) by flooding with oil till initial water saturation (S_{wi}) by the unsteady state JBN method following the

Buckley Leverette assumptions (Latil 1980) The core was again flooded with secondary brine to measure the effective permeability of brine (K_{ew}) by flooding with brine till residual oil saturation (S_{or}) again by the unsteady state JBN method following the Buckley Leverette assumptions (Latil 1980) The relative permeabilities (K_{rw} and K_{ro}) were calculated by dividing the effective permeability's with absolute permeability of the core. Details of JBN method is discussed in (Mamudu 2016; Mamudu *et al.* 2017; Shikhov *et al.* 2017; Carrillo *et al.* 2008). The selected formulated slugs SDS + NaOH, BL + NaOH and SDS + BL + NaOH were then flooded to recover the Sor and the calculations were done as per the unsteady state. The effluents collected at the outlet of the coreflood were taken for adsorption study (Table 6).

Table 6. The flooding experiment was conducted by BOBS (Brine Oil Brine Slug) steps.

Primary brine flooding	Oil flooding	Secondary brine flooding	Final slug flooding
$K_a = \frac{Q\mu L}{\Delta P}$	$S_{oi} = \frac{Brineflushout - LV}{PV}$	$S_{or} = \frac{Soi - oilflushout}{PV} \times 100\,\%$	$R_{Sor} = \frac{oil\,recover\,by\,final\,slug}{volume\,of\,initial\,oil} \times 100\,\%$
Where,			
D = Diameter of core, cm L = Length of the core, cm	S_{oi} = Intial oil saturation S_{wc} = Connate water saturation LV = line volume PV = Pore volume	S_{or} = residual oil saturation	R_{Sor} = Recovery of residual oil out of initial oil saturation

The formulas used for core flood:

$$PV = \frac{\Pi}{4} D^2 L \times \Phi \tag{2}$$

2.3 Adsorption Experiment

Static Adsorption Experiment

Stock solutions were prepared as in Table 7 and from the stock solution experimental samples were prepared as in the Table 8. 1 gm of adsorbents (porous media as in Table 1, SN 5) were added to 10 ml of adsorbate (experimental samples as in Table 8) and allowed for a contact time of 1 h in Rotospin, the samples were centrifuged at 3000 rpm for 20 min at the Centrifuge. Wave length of the BL, SDS and BL + SDS with or without alkali NaOH were determined by Spectrum method (Cazedey and Salgado 2012). The quantity of adsorbents adsorbed onto the adsorbents were determined by finding the difference of adsorbents before and after adsorption. The absorbances were determined by the photometric method in the UV-Spectrophotometer instrument. The amount of adsorbate adsorbed on the adsorbent, q_e(mg/g), was calculated by mass balance Eq. (3):

Table 7. Stock samples for Adsorption experimental work

SN	Conc., μg/ml	Vol. of 0.5% of 0.1 M SDS/0.5% of 0.1 M SDS + 0.6% NaOH/0.7% BL/0.7% BL + 0.6% NaOH/0.7%BL + 0.5% SDS + 0.6% NaOH, μl	Density, gm/ml (assume)	Distil Water (DW)
1	10,000	10	1	9.99

Table 8. Preparation of adsorption experimental samples from stock sample

SN	Conc., μg/ml	Vol. taken from stock sample 10,000 μg/ml	DW, ml
1	10	10	9.99
2	20	20	9.98
3	30	30	9.97
4	40	40	9.96
5	50	50	9.95

Adsorption (mg/g):

$$q_s = \frac{V_s(C_i - C_e) \times 10^{-3}}{m} \text{ mg/g} \tag{3}$$

Where,

q_s = Surfactant adsorption on PM, mg/g
C_i = Surfactant concentration in initial solution before equilibrated with PM, mg/L
C_e = Surfactant concentration in aqueous solution after equilibrated with PM, mg/L
M = weight of the PM, g
V_s = Total volume of solution in original bulk solution, ml

Dynamic Adsorption Experiment
The formulated slugs SDS, SDS + NaOH, BL, BL + NaOH, SDS + BL and SDS + BL + NaOH were injected to recover the residual oil saturation at reservoir temperature 75 °C. The effluents collected from at outlet of the coreflood were taken to adsorption study. Both Influents and effluents were diluted to the concentration ranges from 10, 20, 30, 40 and 50 μg/ml. The wavelength of each formulated slug and absorbances were determined by the Spectrum method and photometric method in the UV-Spectrophotometer instrument. Then amount of dynamic adsorption on PM were calculated by the following Eq. (4) (Ahmadi and Shadizadeh 2015).

$$q_d = \frac{(C_i - C_e) \times V_d}{W} \tag{4}$$

Where,

q_d = Surfactant adsorption on PM, mg/g
C_i = Surfactant concentration in initial solution before coreflood, mg/L
C_e = Surfactant concentration at oulet of coreflood, mg/L
W = Mass of the PM, g
V_d = Volume of injected surfactant solution, L

2.4 Adsorption Isotherms

Langmuir isotherm (Gogoi 2011; Ahmadi and Shadizadeh 2015):

The Langmuir theory is typically applicable through considering a primitive assumption that the sorption occurs at special occurs at special homogenous sites in the adsorbent. Furthermore, the rate of adsorption on the PM is a strong function of both driving forces, the concentration in the solution, and area, amount of bare surface. The solid-phase adsorbate concentration q_e or known as the uptake is mathematically and isothermally related to the equilibrium liquid concentration isotherm with the Langmuir equation, proposed by Irving Langmuir (1916) as the following:

Nonlinear from:

$$q_e = \frac{q_o K_{ad} C_e}{1 + K_{ad} C_e}$$

In linear from:

$$\frac{1}{q_e} = \frac{1}{q_0} + \frac{1}{q_0 K_{ad} C_e} \tag{5}$$

Hence, a graph of $1/q_e$ vs. $1/C_e$ ought to present a straight line whether Langmuir isotherm is followed by the adsorption equilibrium. The slope and intercept of the drawn line are used to calculate the amounts of q_0 and K_{ad}, respectively.

Freundlich isotherm (Gogoi 2011; Ahmadi and Shadizadeh 2015):

An empirical model to depict the adsorption process was developed by Frendlich (1906). An adsorbent with a heterogeneous surface has a variety of different adsorption sites, it is fundamental surface has a variety of different adsorption of his theory. He proved that the "amount of solute adsorbed onto a given mass of an adsorbent to the concentration of the solute in the solute in the solution" as general form of ratio changes vs. solution in the solution. The mathematics is gained to present of multilayer sorption of the surface.

Nonlinear from:

$$q_e = K_f C_e^{1/n}$$

Linear from:

$$\log q_e = \log K_f + \frac{1}{n} \log C_e \tag{6}$$

Where, K_f = Freundlich adsorption constants, mg/g and n = sorption intensity.
Temkin isotherm (Ahmadi and Shadizadeh 2015):
This isotherm contains a factor that explicitly taking into the account of adsorbent–adsorbate interactions. By ignoring the extremely low and large value of concentrations, the model assumes that heat of adsorption (function of temperature) of all molecules in the layer would decrease linearly rather than logarithmic with coverage.

Nonlinear from:

$$q_e = \frac{RT}{b} \ln(K_T C_e), \text{ mg/g} \tag{7}$$

Linear from:

$$q_e = \frac{RT}{b} \ln K_T + \frac{RT}{b} \ln c_e$$

Putting, $B = \frac{RT}{b}$, Eq. (7) becomes $q_e = B\ln(K_T) + B\ln(c_e)$

Where, K_T = Temkin isotherm equilibrium constant, (L/g); b_T = Temkin isotherm constant; R = Universal gas constant (8.3145 J/mol/K); T = Temperature at 298°K; B = Constant related to heat of adsorption (J/mol).
Linear isotherm (Ahmadi and Shadizadeh 2015):
The most basic form of chromatographic theorem, the adsorption isotherm for the solute has been assumed to be linear. The Henry equation describes as the following.

$$q_e = K_H C_e \tag{8}$$

Where, q_e, C_e and K_H are the amount adsorbed, the concentration and constant respectively in units of L/m^2. Linear isotherm is only the low concentration range where adsorption isotherm is linear. The more concentration in solution, the more non-linear behavior has a convex trend (Fig. 1).

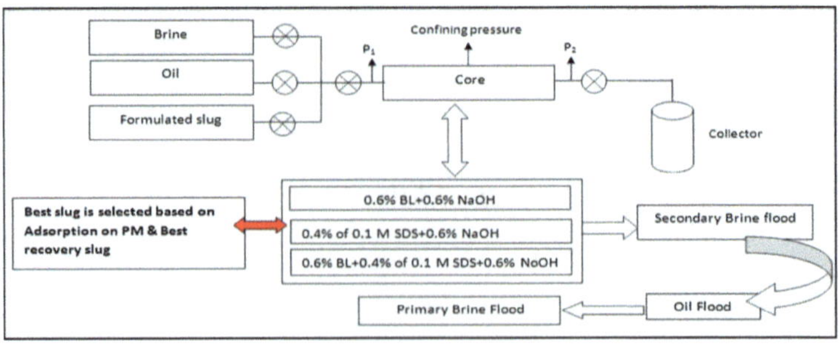

Fig. 1. Graphical abstract

3 Results and Discussion

Acid no. of crude oil

The acid no. of the crude oil sample (Table 1 SN 4) was found to be 1.22 mg of KOH/gm of crude oil, Fig. 2(a) and (b) shows the titration experiment for determining acid number of crude oil. Acid numbers > 1.0 is generally considered high, 0.3–1 is intermediate and 0.1–0.25 is low. The acid numbers of the crude oil under study in this paper was found to be high, which is in the order of 1.22 mg/g. Most crude oils have acid number lower than 5 mg of KOH/g of crude oil. Practically, the minimum acid requirement for crude oil to be effective for ASP flooding is 0.3 mg KOH/g in oil (Chang et al. 2006; Sheng 2015). So from applicability point of view the depleted reservoir with this crude oil can go for alkaline EOR.

(a) (b)

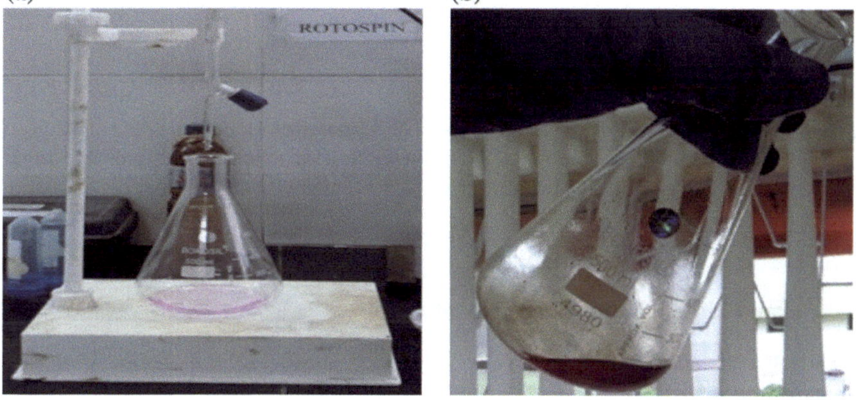

Fig. 2. (a) Purple color indicate the neutral point. (b) Tends to red wine color indicate the present of acid in crude oil.

Identification of acid groups by GC-MS

The presence of carboxylic acid groups in crude oil was identified by GC-MS was shown in Fig. 3(a) to (d). Figure 3(a) showed the GC fingerprints of peak 1, 2, 3 and these peaks are analyzed by Agilent NIST library data 5920 (Agilent GC 7820A and Agilent MS 5920, 2007) and found probable organics for peak 1, 2 & 3 are 3(10)-Caren-4-ol, Hydrocinnamic acid, o-[(1, 2, 3, 4-tetradro-2 naphthyl) methyl]- and -5, 8, 11-Heptadecatriynoic acid respectively. Peak 2 showed the crude oil is acidic and naphthenic base since there were the presences of –COOH- ring structures. Peak 3 also showed the presence of –COOH-. Therefore, from GC-MS it is identified that crude oil is naphthenic base crude oil.

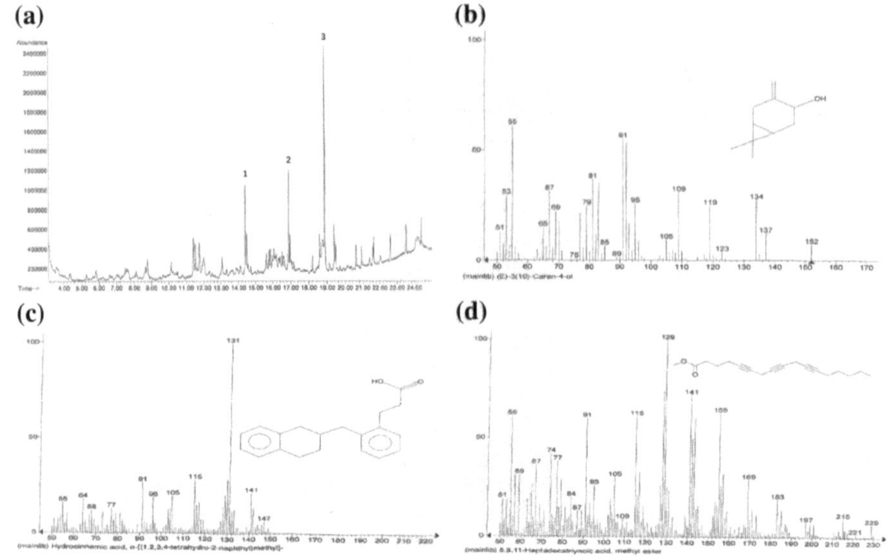

Fig. 3. (a) GC fingerprints of crude oil. (b) -3(10)- Caren-4-ol probable organics identify for peak1. (c) Hydrocinnamic acid, o-[(1, 2, 3, 4-tetradro-2 naphthyl)methyl]-probable organic identify for peak 2. (d) -5, 8, 11-Heptadecatriynoic acid, methyl ester probable organic identify for peak 3

3.1 Formulation of Laboratory S/A Slug

Critical Micelle Concentrations (CMC) for BL and SDS in 3000 ppm Brine Concentration

Gogoi, 2014 (Gogoi 2014) reported that the average salinity of Upper Assam Basin oil reservoirs lies between 3000–5000 ppm. For experimental work we took brine concentration of 3000 ppm salinity to match with the formation water brine salinity. Experiments were done at 75 °C to match the reservoir temperature. Barial formation reservoir temperature was found to be 70–90 °C from the geothermal gradient. Lowest IFT of BL in 3000 ppm brine was 1.8 mN/m at the 0.6% CMC for 0.1 M SDS in 3000 ppm brine was 0.98 mN/m at 0.4% CMC value as shown in Fig. 4. At the CMC of the aqueous phases, emulsions form micelles, the hydrophobic tails form the core of the micelle and hydrophilic heads are in contact with the crude oil, thereby reduce IFT of aqueous phase.

The CMC is an important characteristic of a surfactant. Prior to reaching the CMC, the IFT changes strongly with the concentration of the surfactant. After reaching the CMC, the IFT remains relatively constant or changes with a lower slope. The value of the CMC for a given aqueous and oleic phases is dependent on temperature, pressure, strongly on the presence and concentration of other surface-active substances which in our case were SDS, BL and to a certain extent NaOH. Micelles are formed above CMC value (Das et al. 2017). Anionic surfactants are extremely sensitive to any type salinity, hardness i.e. Ca^{2+}, Mg^{2+} present in the reservoir aqueous phase (Touhami et al. 1998; Ding et al. 2016). Gogoi (2014) reported as the salinity of the aqueous phase increases there is reduction in IFT as compared to deionized water.

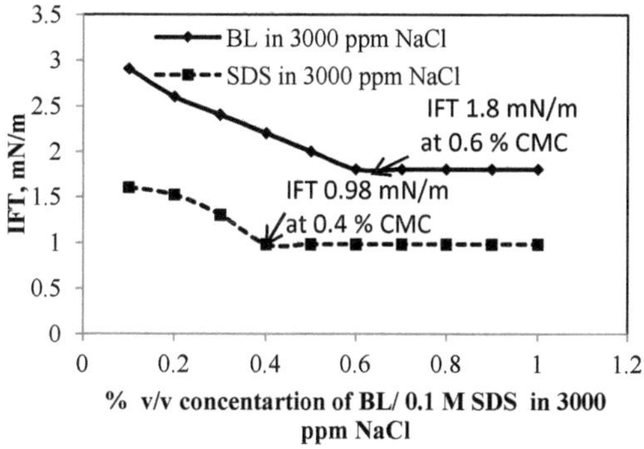

Fig. 4. IFT vs Concentration (%v/v) of BL/SDS in 3000 ppm

Influence of Alkali Interact with Acidic Crude Oil and Their Effectiveness on IFT
Two slugs were formulated for EOR, where the CMC of the surfactants as in Fig. 4 were taken. To the CMC value of the surfactants, either SDS or BL, different concentration of NaOH were added as in Table 9. From Tables 9 and 10 it was found that the reduction of IFT on addition of NaOH in case of BL 93.89%, SDS 95.20% and mixed slug BL + SDS was 96.57%. Therefore, the slug comprising of mixed 0.6% (v/v) BL + 0.4% (v/v) of 0.1 M SDS + 0.6% (w/v) of NaOH is a better slug for EOR than the slug comprising of CMC value of 0.6% (v/v) BL and 0.6% (w/v) of NaOH and 0.4% (v/v) of 0.1 M SDS and 0.6% (w/v) of NaOH. All the measurements were conducted at reservoir temperature of 75 °C. The presence of carboxylic group (-COOH) in acidic crude reacts with NaOH through the process of saponification leading to in situ formation of surfactants, which in turn decreases IFT between oil and water by a mixed interfacial layer and mixed micelles formed by ionized and un-

Table 9. Effect of Alkali and percentage of IFT reduction taking as reference IFT value at CMC

SN	Conc. of NaOH, w/v %	0.6% BL in 3000 ppm		0.4% 0.1 M SDS in 3000 ppm	
		IFT, mN/m	% of reduction of IFT by NaOH	IFT, mN/m	% of reduction of IFT by NaOH
1	0	1.8	0	0.98	0
2	0.1	1.62	10	0.71	27.55102
3	0.2	1.32	26.66667	0.32	67.34694
4	0.3	1.11	38.33333	0.098	90
5	0.4	0.98	45.55556	0.079	91.93878
6	0.5	0.54	70	0.065	93.36735
7	0.6	0.11	93.88889	0.047	95.20408

Table 10. Effect on IFT by Mixing the CMC of SDS and BL

SN	Mixed surfactants	IFT with NaOH	IFT with 0.6% Alkali	% of reduction of IFT by NaOH
1	0.6% BL + 0.4% of 0.1 M SDS in 3000 ppm	1.02 mN/m	0.035	96.57

ionized acids. Hirasaki (2011) observed that an alkaline/surfactant system is a pseudo-two-surfactant system featuring the injection of both surfactant and alkali which saponifies the acidic component present in crude oil thereby generation of in situ surfactants which lowers the IFT between the aqueous and oleic phases and leads to EOR.

The mechanism of in-situ surfactants generation is given below and schematic in Fig. 5. It is assumed that the mixture of acidic species in crude leading to the formation of surfactants is single components HA. Where, HA is the pseudo-acid component and A^- is the surfactant generated. Subscripts w and o stand for aqueous and oil phases. The acid extracted by aqueous phase dissociates to form the hydronium ion (H_3O^+) and surface active anion (A^-). Addition of NaOH reduces the concentration of the H_3O^+ and carries forward dissociation of the acid. Large amount of Na^+ undissociated salt NaA is formed which is oil soluble. On partitioning to the aqueous phase to the aqueous phase the salt is assumed to dissociate completely to produce active species A^- (Borwankar and Wasan 1986; Thomas et al. 2016).

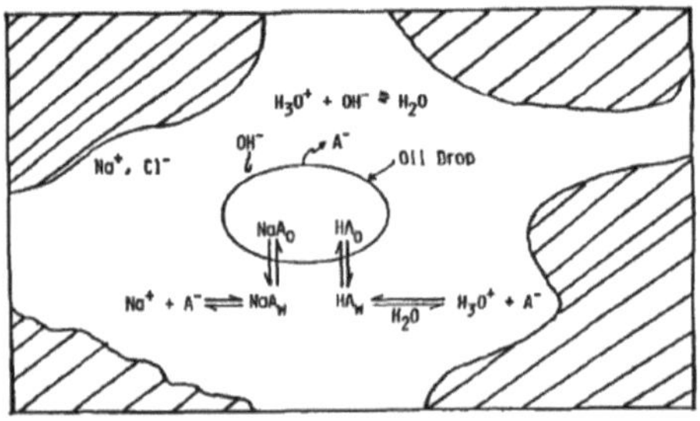

Fig. 5. In-situ surfactant generation by NaOH and acid crude (Borwankar and Wasan 1986)

Surfactant can also increase Capillary number (Nc) by lowering IFT to low or ultralow levels (i.e. 0.01–0.001 mN/m) (Iglauer et al. 2010; Abrams 1975; Green and Willhite 1998). Since NaOH further reduces the IFT from 1.8–0.11 mN/m, 0.98–0.047 mN/m 1.02–0.035 mN/m for BL, SDS and SDS + BL slugs respectively it can be concluded that the Nc increases. Three the slugs under study lowers IFT to levels 10^{-2} to 10^{-3} levels, which will lead to increase of Nc favorable for EOR.

XRD Study

The porous media was characterized by XRD study to determine the types of clay present. In Fig. 6 peaks are single headed peak indicating no impurities in the sample which means the presence of a single phase. The peaks were obtained at 21°, 22°, 24°, 25°, 27°, 28°, 39°, 50°, 60° to 70°. But characteristics picks were obtained at 21°, 27° and 28°. These 2θ values were determined and was found that the presence of mostly Illite clay in porous media (Tucker 1998). Besides, there were peaks of other clay types which include; kaolinite, montmorillonite and chlorite too. Several authors have attempted to address the effect of clay minerals presence on the adsorption of anionic surfactant with single adsorbent (Li et al. 2003; Lv 2011) or clay fraction effect (Amirianshoja et al. 2013). They reported that the kaolinite mineral has the highest adsorption (Johannesen 2007), the impact of mineral surface area also affect adsorption, even with the negative ions (Sánchez-Martín et al. 2008). It was investigated the effect of clay minerals structure and surfactant nature on adsorption, very few researchers have studied the effect of montmorillonite and illite as single adsorbent on the anionic surfactant (Sánchez-Martín et al. 2008; Hower 1970).

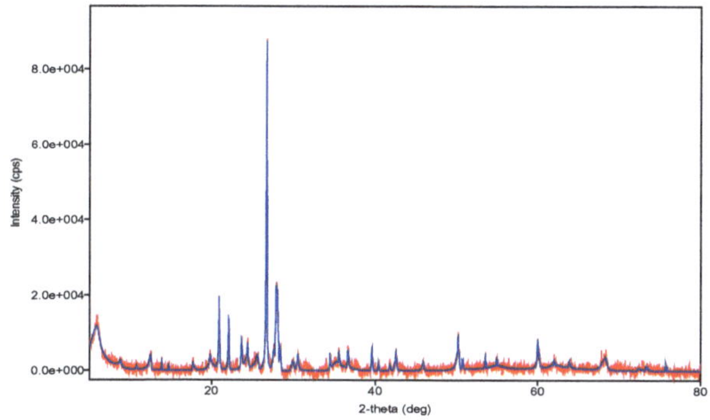

Fig. 6. XRD study of the sand particles in RidakuUltima IV

Core Flooding

Core flood experiments were carried out according to the sequence summarized by Johnson, Bossler and Naumann (JBN) techniques based on Buckley Leverette assumptions (Latil 1980). The reservoir cores were collected from a depth of 3536 m and their average porosity value is 22.61%, respectively. Synthetic brine concentration of 3000 ppm NaCl was used for primary and secondary flooding in core flood experiments and synthetic oil liquid paraffin used for oil flooding in order to meet the compositions of formation water that exists in the reservoir. The Table 11 showed the recovery of each surfactant slug out of initial oil in place (IOIP). The mixed slug BL + SDS + 0.6% NaOH was recovered more about to 32.04% (PV 4.87 cc) of IOIP (15.2 cc) as compared to the BL + NaOH (PV 3.9 cc) and SDS + NaOH (PV 4.76 cc).

Table 11. Total flood volumes for Secondary and Tertiary recovery

Reservoir core (PV 17.94 cc)				
Initial oil saturation, S_{oi} 84.73% (PV 15.2 cc) + Irreducible water saturation, S_{wc} (PV 2.74 cc)				
Primary + Secondary recovery from initial oil saturation, S_{oi} 84.73% of OOIP, (PV 15.2 cc)		Formulated Slug	Tertiary recovery from residual oil 46.05% of IOIP, (PV 7 cc)	
			Recovered oil, % IOIP	Unrecovered, % IOIP
Recovered oil, % IOIP 53.95% (PV 8.2 cc)	Residual oil, % IOIP 46.05% (PV 7 cc)	BL + 0.6% NaOH	25.66% (PV 3.9 cc)	20.39% (PV 3.1 cc)
		SDS + 0.6% NaOH	31.3% (PV 4.76 cc)	14.74% (PV 2.24 cc)
		BL +SDS + 0.6% NaOH	32.04% (PV 4.87 cc)	14.01% (PV 2.13 cc)

Adsorption Isotherms of Slugs with NaOH on PM

The adsorbed amount is expressed in mg of adsorbate weight per gram of adsorbent. The adsorption isotherms were measured at room temperature and atmospheric pressure. The Langmuir, Freundlich, Temkin and Linear isotherms were used to correlate the experimental adsorption data. The below Fig. 7(a, b, c, d, e & f) were the calibration curves, required to found out the unknown concentration of the surfactants after static and dynamic adsorption.

The Fig. 8(a, b & c) and (e, f & g) shows the static and dynamic adsorption (mg/g) vs equilibrium concentration (μg/ml) of the 0.4% of 0.1 M SDS and 0.4% of 0.1 M SDS + 0.6% NaOH, 0.6% BL and 0.6% BL +0.6% NaOH and 0.4% of 0.1 M SDS + 0.6% BL and 0.4% of 0.1 M SDS + 0.6% BL + 0.6% NaOH. It was clearly seen that adsorption of SDS or BL or BL + SDS in 3000 ppm is higher than SDS or BL or BL + SDS with 0.6% NaOH in 3000 ppm in case of both static and dynamic adsorption. That reveals that NaOH reduced the adsorption of surfactants in PM. Adsorption of anionic SDS or BL on the PM was less compared to cationic and nonionic surfactants due to the negativity of clays charged present in reservoir, thus repel each other (Gogoi 2011). With the increase of pH, the adsorption of SDS or BL on the porous media has decreased due to increase of negativity charges in the reservoir (Gogoi 2011).

By comparison between static and dynamic conditions from the Table 12 the adsorption density of mixed slug is much lower than static adsorption (Ahmadi and Shadizadeh 2015). The higher concentration of surfactant the more capacity of surfactant adsorbed on the PM. This was because of the increment in the concentration gradient between both bulk and surface of the PM with the growth in basic concentration of the surfactant solution (Ahmadi and Shadizadeh 2015; Gogoi 2011). The least dynamic adsorption and best recovery among three final slugs found to be in case of the mixed slug 0.6% (v/v) BL + 0.4% (v/v) of 0.1 M SDS + 0.6% (w/v) NaOH. The residual oil was recovered 32.04% (PV 4.87 cc) out of IOIP (PV 15.2 cc) and overall recovery was found to be 85.99% (PV 13.07 cc) out of IOIP (PV 15.2 cc).

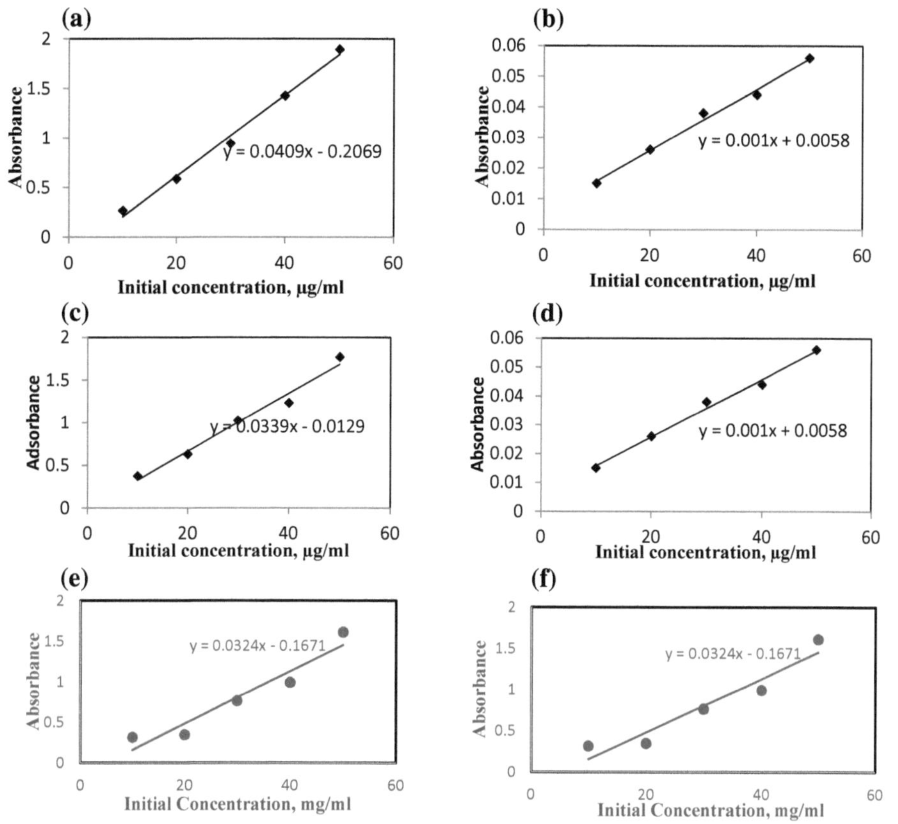

Fig. 7. (a) Calibration curve of 0.4% of 0.1 M SDS in 3000 ppm. (b) Calibration curve of 0.4% of 0.1 M SDS + 0.6% NaOH in 3000 ppm. (c) Calibration curve of 0.6% BL in 3000 ppm. (d) Calibration curve of 0.6%BL + 0.6% NaOH in 3000 ppm. (e) Calibration curve of 0.6% BL + 0.4% of 0.1 M SDS in 3000 ppm. (f) Calibration curve of 0.6% BL + 0.4% of 0.1 M SDS +0.6% NaOH in 3000 ppm

Adsorption Isotherms

The below adsorption isotherms were shown the best fit model for best slug 0.6% BL + 0.4% of 0.1 M SDS +O.6% NaOH in 3000 ppm.

The Fig. 9(a, b, c & d) were shown the four two unknown parameter isotherms for adsorption study. The isotherms were Langmuir, Freundlich, Temkin and Linear isotherms were plotted and found Freundlich was the best fit isotherm for the mixed slug 0.6% BL + 0.4% of 0.1 M SDS +0.6% NaOH. That indicated that the mixed slug was adsorbed on heterogeneous surface by multilayer adsorption (Bera et al. 2013). The Freundlich constant (1/n) is related to the adsorption intensity of the adsorbent. When, $0.1 < 1/n < 0.5$, Adsorption is favorable; $0.5 < 1/n \leq 1$, It is easy to adsorb; $1/n > 1$, It is difficult to adsorb (Bera et al. 2013; Samjey and Dargahi 2010). The Table 13 showed the correlations of each of four isotherms and their parameters. 1/n value for

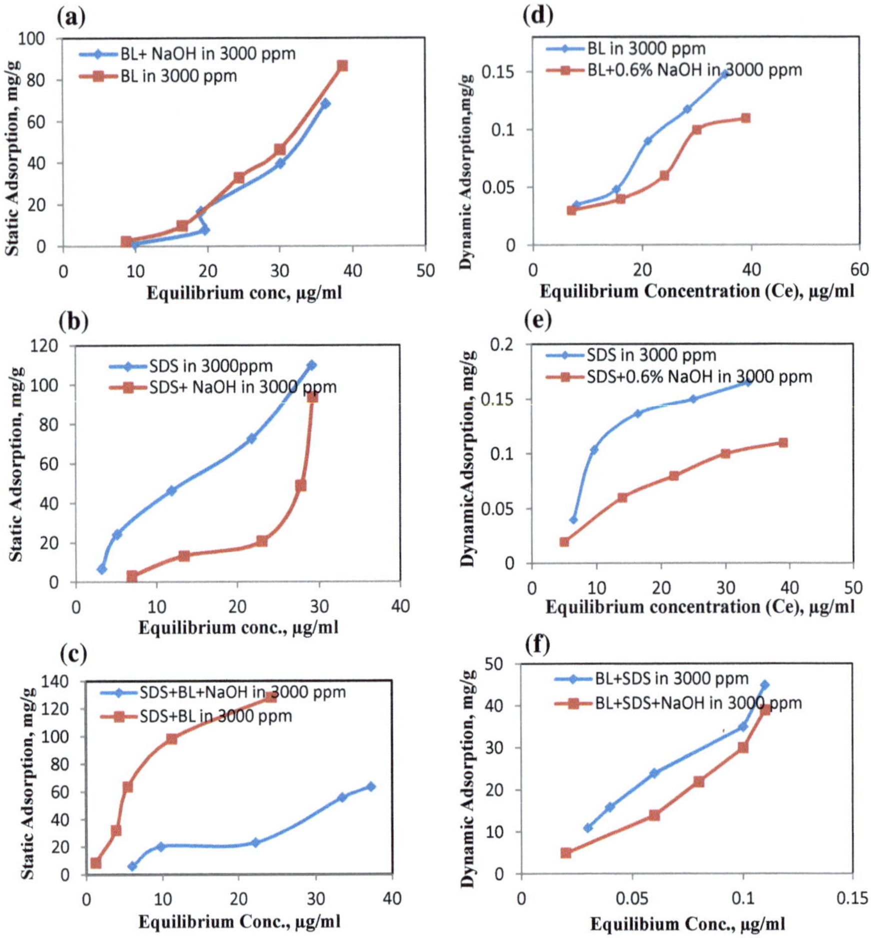

Fig. 8. (a) Static Adsorption (mg/g) vs equilibrium concentration (µg/ml) for BL and 0.6% BL + O.6% NaOH in 3000 ppm. (b) Static Adsorption (mg/g) vs equilibrium concentration (µg/ml) for SDS and 0.4% of 0.1 M SDS +O.6% NaOH in 3000 ppm. (c) Static Adsorption (mg/g) vs equilibrium concentration (µg/ml) for 0.6% BL + 0.4% of 0.1 M SDS and 0.6% BL + 0.4% of 0.1 M SDS +O.6% NaOH in 3000 ppm. (d) Dynamic Adsorption (mg/g) vs equilibrium concentration (µg/ml) for BL and 0.6% BL + O.6% NaOH in 3000 ppm. (e) Dynamic Adsorption (mg/g) vs equilibrium concentration (µg/ml) for SDS and 0.4% of 0.1 M SDS +O.6% NaOH in 3000 ppm. (f) Dynamic Adsorption (mg/g) vs equilibrium concentration (µg/ml) for 0.6% BL + 0.4% of 0.1 M SDS and 0.6% BL + 0.4% of 0.1 M SDS +O.6% NaOH in 3000 ppm

Table 12. Data for static and dynamic adsorption

C_i, µg/ml	Slug	Static adsorption				Dynamic adsorption			
		C_e, µg/ml	q_s, mg/g without NaOH	C_e, µg/ml	q_s, mg/g with NaOH	C_e, µg/ml	q_d, mg/g without NaOH	C_e, µg/ml	q_d, mg/g with NaOH
10	BL or BL + NaOH	8.666	2.333	9.875	1.242	7.909	0.035	7. 091	0.031
20		16.444	9.811	19.557	7.853	15.181	0.048	16.181	0.042
30		24.398	32.888	19.037	16.805	21.321	0.090	24.431	0.060
40		30.066	46.444	30.088	39.732	28.242	0.117	30.127	0.101
50		38.666	86.666	36.277	68.614	35.212	0.147	39.121	0.110
10	SDS or SDS + NaOH	3.16216	6.837	6.882	3.117	6.414	0.040	5. 634	0.021
20		5.05405	24.486	13.352	13.294	9.634	0.104	14. 146	0.062
30		11.8108	46.459	23.058	20.823	16.341	0.136	22. 463	0.081
40		21.8108	72.756	27.764	48.941	25.111	0.151	30.124	0.140
50		29.1081	110.059	29.235	93.823	33.487	0.165	39. 805	0.115
10	BL + SDS or BL + SDS + NaOH	1.210	8.789	6. 526	6.511	11.001	0.030	5.001	0.019
20		3.842	32.315	9.750	20.502	16.201	0.041	14.701	0.060
30		5.421	63.736	22.250	23.250	24.110	0.063	22.321	0.079
40		11.210	98.157	33.501	55.901	35.127	0.107	30.009	0.103
50		24.368	128.157	37.250	63.750	45.013	0.110	39.271	0.119

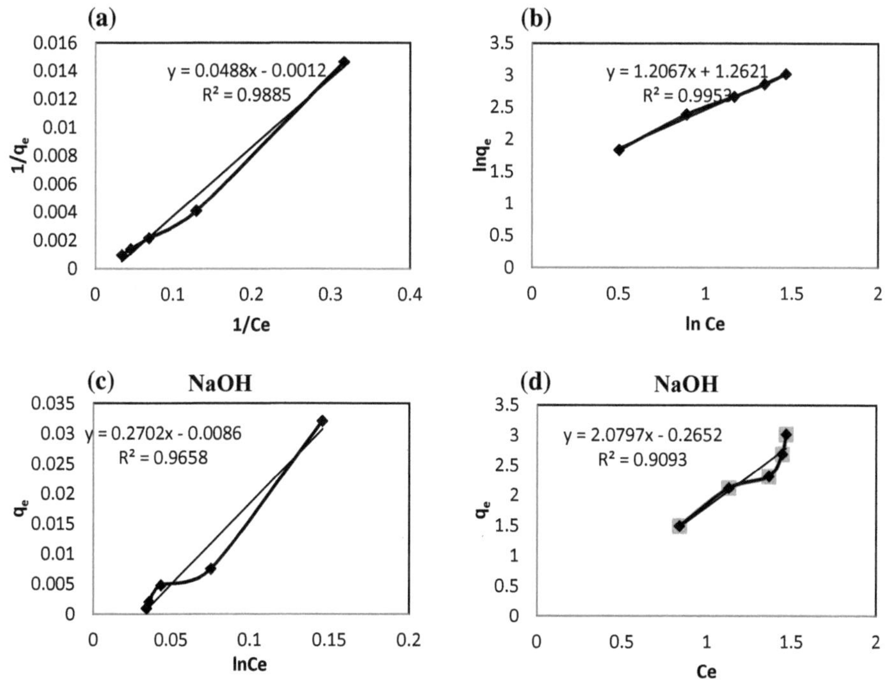

Fig. 9. (a) Langmuir Isotherm model for 0.6% BL + 0.4% of 0.1 M SDS +O.6% NaOH. (b) Freundlich Isotherm model for 0.6% BL + 0.4% of 0.1 M SDS +O.6% NaOH. (c) Temkin Isotherm model for 0.6% BL + 0.4% of 0.1 M SDS +O.6% NaOH. (d) Linear Isotherm model for 0.6% BL + 0.4% of 0.1 M SDS +O.6% NaOH.

Table 13. Parameters of the adsorption isotherms utilized for Slug 0.6% BL + 0.4% of 0.1 M SDS + O.6% NaOH in 3000 ppm

SN	Isotherm	y = mx + c	Correlation	R^2	Parameters		Best fit isotherm
1	Langmuir	y = 0.048x-0.001	$1/q_e = 0.048/C_e$-0.001	0.988	$q_0 \times 10^2$	$K_{ad} \times 10^2$	Freundlich Isotherm
					10	0.48	
2	Freundlich	y = 1.206x+1.262	$\ln q_e = 1.206 \ln C_e + 1.262$	0.996	n	K_f	
					0.829	3.532	
3	Temkin	y = 0.27x0-0.008	$q_e = 0.270 \ln C_e$-0.008	0.965	B	K_t	
					0.270	1.030	
4	Linear	y = 2.079x-0.265	$q_e = 2.079C_e$-0.265	0.909	K_H	C	
					2.079	0.26	

Freundlich obtained 1.206 that means according to the authors Bera et al. 2013 it is difficult to absorbed on PM.

4 Conclusion

The above research works came to the following conclusions:

a. The acid no. of the crude oil sample was found to be 1.22 mg of KOH/gm of crude oil indicated gone for ASEOR flooding.
b. GC-MS was identified that crude oil is naphthenic base crude oil.
c. The CMC of BL and 0.1 M SDS in 3000 ppm brine were obtained at 0.6 v/v % with IFT 1.8 mN/m and 0.4% v/v % with IFT 0.98 mN/m. IFT of both mixed CMC concentration slug (0.6% BL + 0.4% of 0.1 M SDS) was found to be 1.202 mN/m.
d. The percentage reduction of IFT on addition of NaOH in case of BL 93.89%, SDS 95.20% and mixed slug BL + SDS was 96.57%. Therefore, the mixed slug 0.6 (v/v) % BL + 0.4 (v/v) % of 0.1 M SDS + 0.6 (w/v) % NaOH had less IFT slug for EOR than the slugs 0.6 (v/v) % BL + 0.6 (w/v) % NaOH and 0.4 (v/v) % of 0.1 M SDS + 0.6 (w/v) % NaOH.
e. The XRD analysis was found that the presence of mostly Illite clay in porous media.
f. The mixed slug 0.6 (v/v) % BL + 0.4 (v/v) % of 0.1 M SDS + 0.6 (w/v) % NaOH was recovered more about to 32.04% (PV 4.87 cc) of IOIP (15.2 cc) as compared to the 0.6 (v/v) % BL + 0.6 (w/v) % NaOH (25.55%, PV 3.9 cc) and 0.4 (v/v) % of 0.1 M SDS + 0.6 (w/v) % NaOH (31.3%, PV 4.76 cc).
g. Experiments were done to generate the required data to identify the associated parameters of the Langmuir, Freundlich, Temkin and Linear isotherms. The model Freundlich was found to be best fit isotherm for the final slug 0.6 (v/v) % BL + 0.4 (v/v) % of 0.1 M SDS + 0.6 (w/v) % NaOH.
h. The adsorption in dynamic condition was found to be less than the static condition.
i. The more surfactant concentration, the more surfactant adsorption on PM. It was true until reaching the saturation point.
j. The addition of alkali to the surfactant reduced the adsorption of surfactant on the PM.

k. Therefore, the mixed slug 0.6 (v/v) % BL + 0.4 (v/v) % of 0.1 M SDS + 0.6 (w/v) % NaOH was found to be better slug for ASEOR in terms of adsorption and residual oil recovery.

Acknowledgement. We would like to acknowledge the Indo-US Project F. No. 194-1/2009(IC) dated February, 2015 entitled "Foam-assisted CO_2 Flooding for the depleted Reservoirs of Upper Assam Basin and in Candidate Reservoirs in Louisiana" of 21st Century Knowledge Initiative Program under University Grants Commission to carry out the all experimental works.

References

Gogoi, S.B.: Revisiting Naharkatiya core sample for enhanced oil recovery. Indian J. Petrol. Geol. **18**(1), 51–61 (2009)

Gogoi, S.B.: Evaluation of lignin based surfactants for enhanced oil recovery of Naharkatiya porous media. Indian Chem. Eng. J. **50**(1), 47–55 (2010)

Shen, P., Yu, J.: Fundamental Research on Enhanced Oil Recovery in Large Scale. Petroleum Industry Publication Company, Beijing (2002)

Wang, D., Cheng, J., Li, Q., Jiang, Y., Sun, Y., He, Y.: Experience of IOR practices from large-scale implementation in layered sandstones. Paper SPE 64287 presented at the SPE Asia Pacific Oil and Gas Conference and Exhibition, Brisbane, Australia, 16–18 October 2000 (2000). https://doi.org/10.2118/64287-ms

Wang, D., Sun, Y., Wang, Y., Tang, X.: Producing more than 75% of Daqing OilField's production by IOR, what experiences have been learnt? Paper SPE77871 presented at the SPE Asia Pacific Oil and Gas Conference and Exhibition, Melbourne, Australia, 8–10 October 2002 (2002). https://doi.org/10.2118/77871-ms

Sui, J., Yang, C.-Z., Yang, Z.-Y., et al.: Surfactant-alkaline-polymer FloodingPilot project in non-acidic paraffin oil field in Daqing. Paper SPE 64509 presented at the SPE Asia Pacific Oil and Gas Conference and Exhibition, Brisbane, Australia, 16–18 October 2000 (2000). https://doi.org/10.2118/64509-ms

Arihara, N., Yoneyama, T., Akita, Y., Xiang Guo, L.: Oil recovery mechanisms of alkali-surfactant-polymer flooding. Paper SPE 54330 presented at the SPE Asia Pacific Oil and Gas Conference and Exhibition, Jakarta, 20–22 April 1999 (1999). https://doi.org/10.2118/54330-ms

Guo, W.: Status and development direction of tertiary oil recovery technique in Daqing Oilfield. Petrol. Geol. Prod. Daqing Oilfield **18**(3), 24–26 (1999)

NIST library data from Agilent GC-7820A and Agilent MS 5920 (2007) Version

Hurst, A.: Textural and geochemical micro-analysis in the interpretation of clay mineral characteristics, lessons from sandstone hydrocarbon reservoirs. Clay Miner. **34**, 137–149 (1999)

Tucker, M.: Techniques in Sedimentology, pp. 198–203. Blackwell Scientific, Oxford, London (1998)

Cazedey, E.C.L., Salgado, H.R.N.: Spectrophotometric determination of ciprofloxacin hydrochloride in ophthalmic solution. Adv. Anal. Chem. **2**(6), 74–79 (2012)

Gogoi, S.B.: Adsorption–desorption of surfactant for enhanced oil recovery. Transp. Porous Media **90**(2), 589–604 (2011)

Ahmadi, M.A., Shadizadeh, S.R.: Experimental investigation of a natural surfactant adsorption on shale–sandstone reservoir rocks: static and dynamic conditions. Fuel **159**, 15–26 (2015)

Chang, H.L., Zhang, Z.Q., Wang, Q.M., Xu, Z.S., Guo, Z.D., Sun, H.Q., Cao, K.L., Xiao, Q.: Advances in polymer flooding and alkaline/surfactant/polymer processes as developed and applied in the People's Republic of China. JPT **58**, 84–89 (2006)

Sheng, J.J.: Investigation of alkaline–crude oil reaction. Petroleum **1**(1), 31–39 (2015)

Gogoi, S.B.: Effluent as surfactant for enhanced oil recovery. Innovative Energy Policies **3**(1), 46–54 (2014)

Das, B.M., Gogoi, S.B., Mech, D.: Micellar-polymer for enhanced oil recovery for upper Assam basin. Resour. Efficient Technol. J. **3**, 82–87 (2017)

Latil, M.: Enhanced oil recovery, Institut Francais DU Petrole, Editions TECHNIP 27, Rue Ginux 75737 Paris CEDEX 15, pp. 47–50 (1980)

Touhami, Y., Hornof, V., Neale, G.H.: Dynamic interfacial tension behavior of acidified oil/surfactant-enhanced alkaline systems 2. Theoretical studies. J. Colloids Surf. A Physicochem. Eng. Aspects **133**(3), 211–231 (1998)

Ding, L., Zhang, G., Behling, J., Salinas, J.L.L., Ge, J., Puerto, M.C., Hirasaki, G.J., Miller, C.A.: Determination of active soap number of crude oil and soap portioning behavior. J. Energy Fuel (2016). https://doi.org/10.1021/acs.energyfuels.6bo1603

Hirasaki, G.J., Miller, C.A., Puerto, M.: Recent Advances in Surfactant EOR, Rice University, SPE 115386 (2011)

Borwankar, R.P., Wasan, D.T.: Dynamic IFT in acidic crude oil/caustic systems. AIChE J. **32**(3), 255–466 (1986)

Thomas, N.C., Ghosh, B., AlAmeri, W.S., Kilybay, A.: Alkali and hybrid-alkali flooding as a tertiary oil recovery mode: prospects and challenges. Int. J. Petrol. Petrochem. Eng. (IJPPE) **2** (2), 22–31 (2016)

Iglauer, S., Wu, Y., Shuler, P., Tang, Y., Goddard, W.A.: New surfactant classes for enhanced oil recovery and their tertiary oil recovery potential. J. Petrol. Sci. Eng. **71**, 23–29 (2010)

Abrams, A.: The influence of fluid viscosity, interfacial tension, and flow velocity on residual oil saturation left by waterflood. SPEJ **15**(05), 437–447 (1975)

Green, D.W., Willhite, G.P.: Enhanced Oil Recovery: SPE Textbook Series, vol. 6, 545 p. SPE, Richardson (1998)

Li, H., Sheng, G., Teppen, B.J., Johnston, C.T., Boyd, S.A.: Sorption and desorption of pesticides by clay minerals and humic acid-clay complexes. Soil Sci. Soc. Am. J. **67**(1), 122–131 (2003)

Lv, W.: Static and dynamic adsorption of anionic and amphoteric surfactants with and without the presence of alkali. J. Petrol. Sci. Eng. **77**(2), 209–218 (2011)

Amirianshoja, T., Radzuan, J., Kamal Idris, A., Rahmani, O.: A comparative study of surfactant adsorption by clay minerals. J. Petrol. Sci. Eng. **101**, 21–27 (2013)

Sánchez-Martín, M., Dorado, M.C., del Hoyo, C., Rodríguez-Cruz, M.S.: Influence of clay mineral structure and surfactant nature on the adsorption capacity of surfactants by clays. J. Hazard. Mater. **150**(1), 115–123 (2008)

Hower, W.F.: Adsorption of surfactants on montmorillonite. Clays Clay Miner. **18**(97105), 97–105 (1970)

Johannesen, E.B., Graue, A., Bergen, U.: Systematic investigation of waterflood reducing residual oil saturations by increasing differential pressures at various wettabilities, SPE 108593 (2007)

Bera, A., Kumar, T., Ojha, K., Mandal, A.: Adsorption of surfactants on sand surface in enhanced oil recovery: isotherms, kinetics and thermodynamics studies. Appl. Surf. Sci. **284** (2013), 87–99 (2013)

Samjey, A., Dargahi, M.: Kinetics and thermodynamics of adsorption of Congo red on cellulose. Cent. Eur J. Chem. **8**(2010), 906–912 (2010)

Formulation of Chemical Slug for EOR Application of Moran Oil Field of Upper Assam Basin, India

Joyshree Barman[1], Subrata Borgohain Gogoi[1(✉)], Jayakumar Viswanathan[2], and Debasish Konwar[1]

[1] Department of Petroleum Technology, Dibrugarh University, Dibrugarh 786004, Assam, India
subrata@gmail.com
[2] Petroleum Engineering Department, Universiti Teknologi PETRONAS, 32610 Seri Iskandar, Tronoh, Perak, Malaysia

Abstract. Generally, most of the depleting reservoirs of Upper Assam Basin, India are treated with chemical enhanced oil recovery (CEOR) processes which are one of the tertiary oil recoveries to recover about 30–60% of oil which cannot be produced by means of primary or secondary recovery methods. CEOR processes are mainly emphasise on injecting surfactants (S), alkalis (A) and polymers (P) or a combination S+A or A+P or S+P or A+S+P to the aqueous phase of the depleting reservoirs. The S reduces interfacial tension (IFT) between oil and aqueous phase and increases displacement efficiency (E_D) of the reservoirs. While the A tend to reduce the wettability from oil wet or less water wet to more water wet. A also decrease the adsorption of S into the porous media. Water floods are often very inefficient in naturally sandstone oil reservoirs because many of these reservoirs are mixed-wet or oil-wet towards the end of secondary brine flooding as well as extremely heterogeneous. These reservoirs are challenging targets for chemical flooding because they typically have a high permeability contrast between the fractures and the matrix with low matrix permeability. Some of the world's largest oil reservoirs are fractured sandstone with high reservoir temperature and high salinity formation brine. Some of them also have low API° gravity oils, which increases the difficulty of recovering the oil. A slug comprising of surfactant and alkali has been developed that shows promising results for such difficult reservoirs. Ultra-low interfacial tension (IFT) and good aqueous stability were achieved with this surfactant slug in high salinity brine at a high reservoir temperature of 78 °C. Both static and dynamic imbibition experiments were conducted using a sandstone core of Moran Oil Field. Oil recovery was higher in only SDS+NaOH slug flooding compared to flooding of SDS, NaOH and SDS+NaOH sequentially. The oil recovery is good taking into account that the temperature and salinity conditions were harsh, the core was less permeable and fractured, no mobility control was used, and only a small surfactant slug was injected. The unsteady state coreflood results were interpreted using Johnson-Bossler-Naumann (JBN) method. It showed that both the mechanisms of IFT reduction and wettability alteration were important for oil recovery. Neither IFT reduction nor wettability alteration alone recovered oil as high as the combined contributions from both. So, this study focuses on effect of S, A followed by S+A slug flooding (F1) or flooding of S+A only (F2)

H. Ameen and T. Sorour (Eds.): GeoMEast 2018, SUCI, pp. 137–150, 2019.
https://doi.org/10.1007/978-3-030-01929-7_10

to determine the efficiency of residual oil recovery from the depleted Moran Oil Field. The residual oil recovery was calculated by JBN method based on Buckley Leveret assumptions for unsteady state process.

Keywords: Alkaline (A) · Enhanced oil recovery (EOR)
Interfacial tension (IFT) · Surfactants

Nomenclature

S	Surfactants
A	Alkalies
P	Polymers
CEOR	Chemical Enhanced Oil Recovery
CP	Core plug
ST	Surface tension
IFT	Interfacial tension
HLB	Hydrophile-lipophile balance
ml	millilitre
DW	Distilled water
ppm	parts per million
mN/m	millinewton per meter
wt./V%	weight per volume percentage
$\Delta\rho$ (g/cc)	Density difference in gram per cubic centimetre
E_S	Sweep efficiency
E_D	Displacement efficiency
SDS	0.5 Molar of Sodium Dodecyl Sulfate
S+A	Surfactant alkali
NaOH	1.0 Molar of Sodium Hydroxide
SDS+NaOH	1.6 wt./V% of SDS+2.0 wt./V% of NaOH
EOR	Enhanced Oil Recovery
CMC	Critical Missile Concentration

1 Introduction

Generally, tertiary oil recovery or best known as EOR methods are best applicable for matured oil fields which has been producing for decades. The production of oil from these oil fields has reached its maximum capacity and began to deplete. EOR processes are classified (Das 2012) into thermal, chemical, miscible carbon dioxide (CO_2) and microbial enhanced oil recovery (MEOR). Similarly, thermal EOR processes apply heat to the crude oil (CO) in porous media of formations, reducing the viscosity of CO and increases the relative permeability of oil, K_{ro}. According to (Elias and Ramon 2013), the methods applied are in-situ combustion, steam flooding and cyclic injection as well as steam assisted gravity drainage. Meanwhile, CEOR processes are alternatives of injecting more costly and exotic fluids such as S, A and P or a combination of S+A or A+P or S+P or A+S+P to the aqueous phase of the reservoirs. Thus, a soap like-material is formed as a result when the injected S, A and P reacts with organic acids of

CO. This may decrease the IFT or capillary pressure which hinders the CO from flowing through a reservoir (Hakiki et al. 2014, 2016). Whereas, MEOR incorporates injection of microbes into reservoirs to partially digest long chained hydrocarbon molecules, produce biosurfactants or emit CO_2. However, MEOR techniques are not widely used due to higher cost and the developments are not accepted. EOR techniques are conducted to recover trapped oil in the porous media of the reservoirs which cannot be extracted during conventional phases. Therefore, CEOR processes enhance almost 30–60% of oil recovery (Hobson 1975; Walsh and Lake 2003; Smith 1966; Palo Alto 1999) compared to primary and secondary oil recovery methods, about 20–40% of oil recovery.

(Pillon 2008) has discussed that surfactants are efficient in reducing IFT between oleic and aqueous phase besides, decreasing capillary pressure of CO allowing the oil to flow throughout the reservoir. This lead to higher E_D and improves EOR (Pillon 2008). Moreover, few studies have reported to give promising and more beneficial impacts on recovering more trapped oil from the formations by injecting S + A in reservoirs as reported in precious studies by (Martin et al. 1985; Nelson et al. 1984; Krumine et al. 1982, 1983; Reisberg and Doscher 1956). S injected concentrated at the interface or boundary of oleic and aqueous phase, reducing IFT to an ultra-low IFT which is approximately (0.001×10^{-3} mN/m) as in studies conducted by (Gogoi and Das 2012). However, the A injected tend to reduce the wettability of reservoir from oil wet or less water wet to more water wet. In addition, A also inhibits the adsorption of S (Gurgel et al. 2008) into porous media of reservoirs.

According to (Mandal 2015) study, it can be concluded that chemical flooding can be described in two basic mechanisms which are macroscopic and microscopic displacement efficiencies. Macroscopic displacement efficiency is related to P injections into formations meanwhile, S and A injections to the porous media of the reservoirs deals with microscopic displacement efficiency. As the IFT between aqueous and oleic phase reduces, the emulsification of oil will be easier and EOR is improved as E_D increases (Mandal 2015; Nagalajan and Wasan 1993; Shen et al. 2010).

Therefore, the objective of this study is to formulate a suitable surfactant-alkali slug to improve EOR depleting reservoirs of Moran Oil Field, Upper Assam Basin. SDS surfactant and NaOH as a representative for alkali were used to formulate S+A slug to be injected to the aqueous phase of the Moran reservoirs. Besides, this study also focusses on effect of S, A followed by S+A slug flooding (F1) or flooding of S+A only (F2) to determine the efficiency of residual oil recovery from the depleted Moran Oil Field. Core flooding experiments were done where the oil recovery was calculated based on Johnson-Bossler-Naumann (JBN) methods for unsteady state process.

2 Experiments

2.1 Determining ST and IFT of SDS

About 10 different concentrations ranging from 0.2 wt./V% to 2.0 wt./V% of 50 ml of SDS in DW and 69440 ppm brine were formulated in centrifuge tubes and placed in rotospin rotary mixer for half an hour at 50 rpm. Then, Krüss EasyDyne S Tensiometer was used to measure the ST of formulated different concentrations of SDS in DW and 6940 ppm brine as the Δρ (g/cc) of the SDS and air was calculated and keyed in the EasyDyne S Tensiometer. Next, the formulated different concentrations of SDS in DW and 6940 ppm brine earlier were conducted using Krüss EasyDyne S Tensiometer by adding toluene as a demonstrative for crude oil to measure the IFT of different concentrations of SDS with toluene. Before that, Δρ (g/cc) of different concentrations of SDS and toluene were calculated and keyed in before measuring the IFT. The Du Nüoy ring was adjusted carefully and slowly to measure the accurate ST and IFT values for different concentrations of SDS.

2.2 Determining ST and IFT of NaOH

50 ml of 10 different concentrations of 0.25 wt./V% to 2.5 wt./V% of NaOH were formulated in DW and 6940 ppm of brine before placing the centrifuge tubes in rotospin rotary mixer for half an hour. Δρ (g/cc) between NaOH and air were calculated and keyed in the Krüss EasyDyne S Tensiometer to measure the ST of different concentrations of formulated NaOH. The same steps above are repeated with different concentrations of NaOH by adding toluene to measure IFT in EasyDyne S Tensiometer.

2.3 Determining ST and IFT of S+A

To formulate S+A, the critical missile concentration (CMC) values of both SDS and NaOH in 6940 ppm brine were mixed together with ratio of 1:1 and was placed in rotospin rotary mixer for 30 min. The ST and IFT of the S+A formulated were measured in Krüss EasyDyne S Tensiometer as steps mentioned above.

2.4 Core Flooding Experiments

Pore volume of the core plug (CP) was determined using TPI-219 Teaching Helium Porosimeter before saturated with 6940 ppm of brine, salinity of M#1 formation water. Initially, brine flooding was done until the CP was fully saturated with 6940 ppm brine to calculate absolute permeability, K_a. Next, the CP was flooded with light paraffin oil. Once the first drop of oil collected at the outlet, the relative permeability of oil (K_{ro}) at irreducible water saturation (S_{wir}) was calculated till the CP is fully saturated with paraffin oil. Again, the CP was flooded with secondary brine till the first drop of brine collected to peruse the relative permeability of water (K_{rw}) at residual oil saturation (S_{or}). The secondary brine flooding continued till no more oil was collected. Furthermore, these whole core flooding experiments were conducted at temperature of 78 °C. The residual oil volume in CP was calculated considering inlet and outlet pipe volume. Therefore, CEOR Flooding 1 (F1) and Flooding 2 (F2) were conducted to determine the most efficient oil recovery from M#1 CP where F1 is flooding of SDS, NaOH followed by SDS + NaOH slug correspondingly. Meanwhile, F2 is flooding of SDS + NaOH slug only. Furthermore, the volume of brine, oil and emulsions collected at outlet from F1 and F2 methods were placed in the centrifuge machine for 10 min at

4500 rpm for better and accurate separation of aqueous and oleic phase. Hence, the percentage of residual oil recovery from EOR F1 and F2 techniques were analysed.

3 Results and Discussion

The calculations below are done to determine the HLB number of SDS surfactant
Davies' method:

$$HLB = 7 + [\sum_{i=1}^{m} H_i - (n \times 0.475)]$$

Where
H_i = Value of the ith hydrophilic groups as in Table 1.
M = Number of hydrophilic groups in molecule.
n = Number of lipophilic groups in molecule.

Table 1. Materials used in the experiments

Sn	Sample	Specifications	Source
1	Core plug M#1	Barail/depth 3200 m	Oil E&P
2	Light paraffin oil	Kinematic viscosity at 37.8 ° C is 30 cS	GlaxoSmithKline Pharmaceuticals Ltd., Mumbai
3	Toluene (C_7H_8)	M.W. 92.14 g/mol, density 0.865–0.867 g/cc	Spectrochem Pvt. Ltd., Mumbai
4	Sodium chloride (NaCl)	M.W. 58.44 g/mol	RFCL Ltd.
5	Sodium hydroxide (NaOH)	M.W. 40.0 g/mol	Thermo Electron LLS India Pvt. Ltd.
6	Acetone (C_3H_6O)	M.W. 58.08 g/mol, boiling range 55.5–56.5 °C	SDFCL Ltd.
7	Distilled water	–	Laboratory supply
8	Sodium dodecyl sulfate (SDS)	M.W. 288.37 g/mol	Merck Specialities Pvt. Ltd., Mumbai

Griffin's method:

$$HLB = 20 \times \frac{M_h}{M}$$

Where
M_h = Molecular mass of hydrophilic part.
M = Molecular mass of the molecule.

The Griffin's method can only be used to calculate and identify the HLB number for non-ionic surfactant. As SDS is an anionic surfactant, the HLB number of SDS was calculated using Davies' method the effect of stronger and weaker hydrophilic parts in the molecules are considered. The calculation of HLB for SDS using Davies' equation shown below:

$$HLB = 7 + [\sum_{i=1}^{m} 38.7 - (12 \times 0.475)]$$

$$HLB = 4$$

As shown in Figs. 1, 2, the hydrophilic part of the SDS molecule is the $SO_4^-Na^+$ which is also known as the head of the SDS molecule. The value for the $SO_4^-Na^+$ is 38.7 as can be seen in Table 1 below. Meanwhile, the long chain of carbons is known as tail part of the SDS molecule, which acts as lipophilic part of the SDS molecule. Thus, there are 12 carbon atoms attached to the long chain of lipophilic tail of molecule and each carbon atom gives value of 0.475 as shown in Table 2 below. Hence, HLB number for SDS calculated using Davies' method gives a value of 40 as discussed (Stavroudis 2009). This specifies the SDS is a water solubilizing agent (Fig. 3).

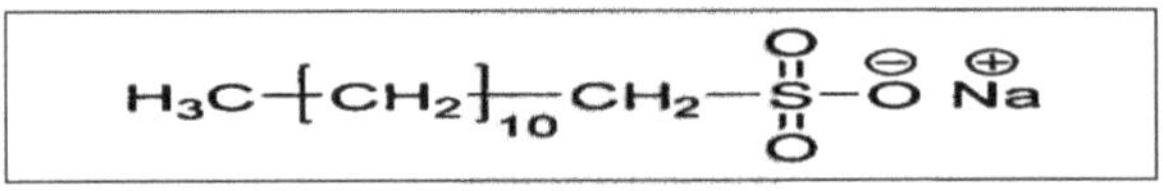

Fig. 1. Chemical structure of SDS

From the graphs of ST and IFT of 0.5 Molar SDS plotted above, it was found to be the ST value between aqueous phase and air and IFT value between oleic and aqueous phase decreases as the salinity of the aqueous phase where SDS was added increases. For example, EasyDyne Tensiometer recorded values of ST = 32.8 mN/m & 26.9 mN/m and IFT = 2.1 mN/m & 1.3 mN/m for 1.0 wt./V% of SDS in DW and 1.0 wt./V of SDS in 6940 ppm of brine respectively. Moreover, an analysis was made as the ST between aqueous phase and air and IFT between oleic and aqueous phase decreases as more concentration of SDS by wt/V% was added to the aqueous phase. After reaching a specific concentration of SDS, the ST and IFT of the SDS were recorded the lowest value and continued constant even the concentration of SDS in wt/V% was increased. Thus, this point was known as SDS CMC where CMC of SDS in DW and CMC of SDS in 6940 ppm of brine were observed to be 1.6 wt./V%. Hence, the aim of this paper was proved as the surfactants reduce the IFT but the surfactants are more efficient in reducing the IFT values between aqueous and oleic phase as the salinity of the aqueous phase increases as shown in Fig. 4.

According to the plotted graphs of ST and IFT of NaOH as shown above, it was found to be the ST and IFT value between oleic and aqueous phase decreases as the salinity of the NaOH aqueous phase increases. That is to say, values of ST = 58.0 mN/m & 42.2 mN/m and IFT = 3.2 mN/m & 2.3 mN/m were recorded for 1.5 wt./V%

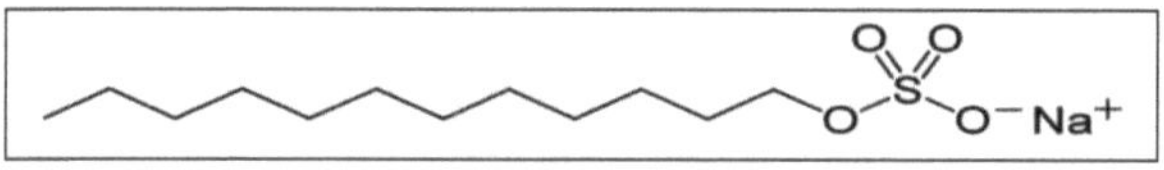

Fig. 2. Skeleton structure of SDS

Table 2. Instruments used in the experiments.

Sn	Instrument	Parameters	Model	Manufacturer
1	Core cutting, plugging, end-facing & grinding machines	Oil recovery	–	Veenedyt, Mahim Mumbai
2	Soxhlet apparatus	Cleaning core	–	Instruments & Equipments, Bombay
3	Ultrasonic cleaner	Cleaning core using 20–400 kHz	500HI	Electrosonic Industries, Mumbai
4	Electronic weighing balance	Weight	A 122	Mettler Toledo, Switzerland
5	Dehumidifier control oven	Humidity	101	Vindish Instruments Pvt. Ltd.
6	Magnetic stirrer	–	Tarsons Digital Spinot	Tarsons Product Pvt. Ltd., India.
7	Rotary rotospin mixer	–	Tarsons 3090X	Tarsons Product Pvt. Ltd., India.
8	Krüss EasyDyne S Tensiometer	ST and IFT	K20S EasyDyne	Krűss GmbH Germany
9	Centrifuge machine	Separating water from oil	Electra Lab Centrifuge	Euro Diagnostic System Pvt. Ltd.
10	Helium porosimeter	Porosity	TPI-219	Coretest Systems Inc., California, USA.
19	Core Flooding System	Oil recovery	–	Gon Engineering Works, Mallick Road, Jharkhand

of NaOH in DW and 1.5 wt./V of NaOH in 6940 ppm of brine correspondingly. Besides, concentration of NaOH (wt/V%) are inversely proportional to ST and IFT between oleic and aqueous phase. As can be seen from the graphs above, the ST and IFT decrease to a constant value as the wt/V% of NaOH increases up to the NaOH CMC value. Therefore, the CMC for NaOH in DW and 6940 ppm brine were determined to be 2.0 wt./V% of NaOH. However, the ST and IFT between aqueous and oleic phase were reduced to the lowest value by adding SDS to the aqueous phase compared to NaOH. For an instance, values of ST = 26.9 mN/m & 46.0 mN/m and

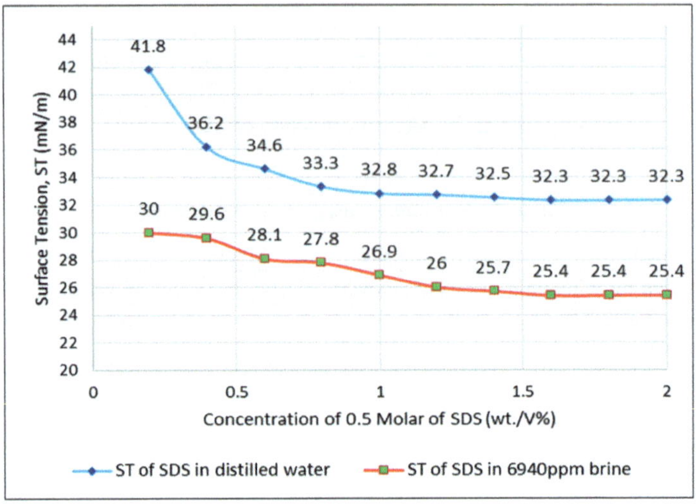

Fig. 3. ST for different concentrations of SDS in DW and 6940 ppm of brine

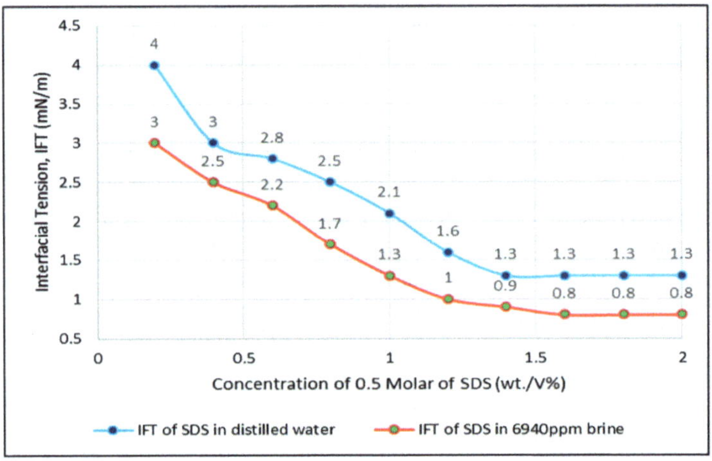

Fig. 4. IFT for different concentrations of SDS in DW and 6940 ppm of brine

IFT = 1.3 mN/m & 4.1 mN/m were observed for 1.0 wt/V% of SDS in 6940 ppm of brine and 1.0 wt/V% of NaOH in 6940 ppm of brine respectively (Figs. 5 and 6).

Therefore, 50 ml of SDS + NaOH slug of 1.6 wt/V% of SDS + 2.0 wt/V% of NaOH in DW and 6940 ppm brine were formulated. EasyDyne Tensiometer measured ST values for the slug formulated to be 40.2 mN/m and 30.5 mN/m for SDS + NaOH formulated in DW and 6940 ppm of brine. Meanwhile, the IFT of the SDS + NaOH slug was further reduced to very low reading of 1.2 mN/m for SDS + NaOH in DW and 0.6 mN/m for SDS + NaOH in 6940 ppm of brine. Hence, the EOR of Moran Oil Field can be most probably improved by this injecting SDS + NaOH to the aqueous

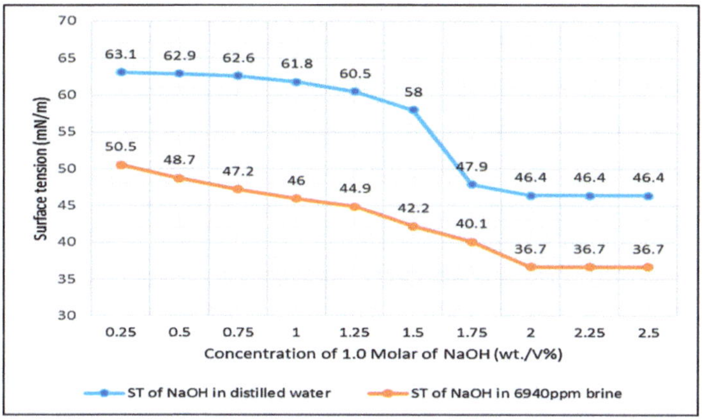

Fig. 5. Graph of ST for different concentrations of NaOH in DW and 6940 ppm brine

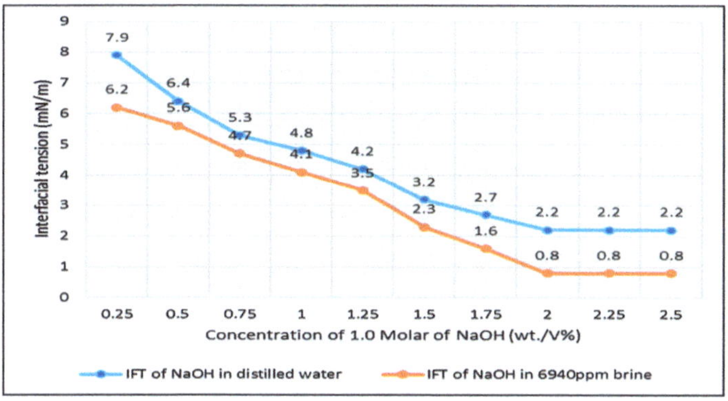

Fig. 6. Graph of IFT for different concentrations of NaOH in DW and 6940 ppm brine

phase of the reservoirs to recover more trapped oil which cannot be produced by means of conventional methods.

Moreover, S_{wir} and S_{or} of M#1 CP was studied to be 17.56% and 77.05% respectively from Fig. 7. Thus, only up to 77.05% from Moran core plug was able to be produced leaving 22.95% of residue oil trapped in the porous media of the formation. Besides, the K_a was studied to be 145.67 mD (0.14567 D) during initial brine flooding. The K_{rw} and K_{ro} during flooding of secondary brine, F1 and F2 CEOR techniques can be analysed as in Figs. 7, 9. According to theoretical knowledge, these relative permeability values are ratio of effective permeability of a specific fluid (oil or water) with respect to the absolute permeability of the core sample.

As can be seen from Figs. 8, 10, F1 CEOR process reported percentage of residual oil recovered with approximately 56.41% meanwhile F2 CEOR technique recovered almost 67.65% of the residual oil volume trapped in the formation which cannot be produced by means of conventional processes, which is quite high compared to

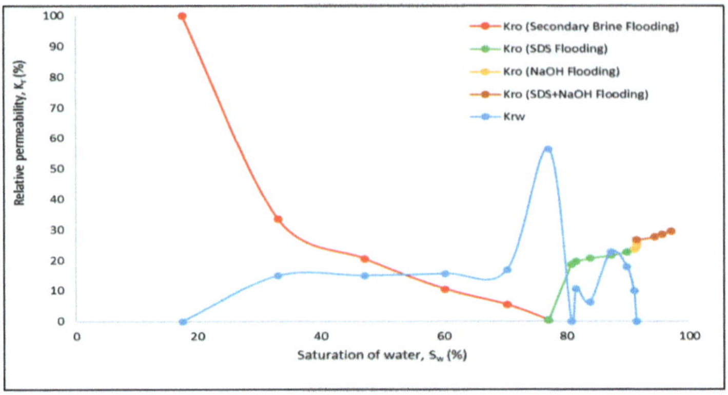

Fig. 7. K_{ro} and K_{rw} in M#1 CP during secondary brine and F1 CEOR technique

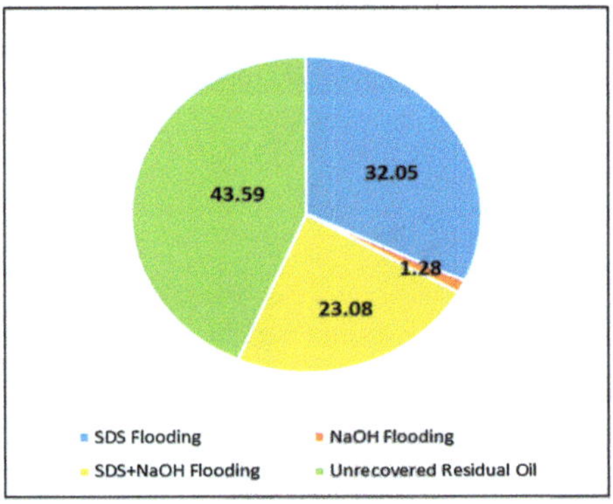

Fig. 8. Residual oil recovery (%) from M#1 core by F1 CEOR technique

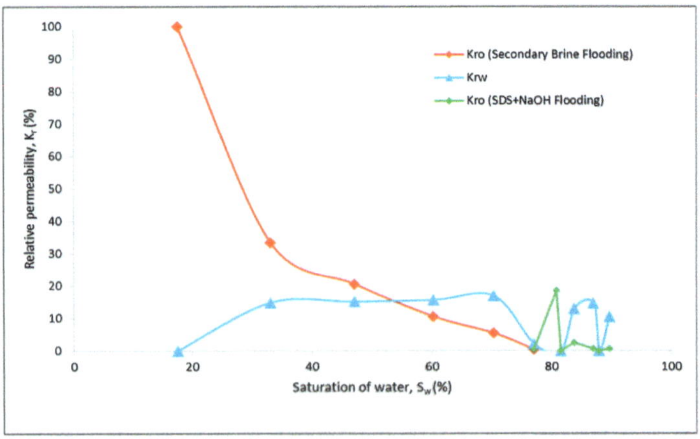

Fig. 9. K_{ro} and K_{rw} in M#1 CP during secondary brine and F2 CEOR technique

F1 CEOR flooding. These results indicate F2 CEOR technique which is flooding of only SDS + NaOH slug recovered about 11.24% of more residual oil compared to F1 CEOR technique, flooding of SDS, then NaOH followed with SDS + NaOH slug. Hence, less volume of residual oil left unrecovered in M#1 CP compared to F1 CEOR technique.

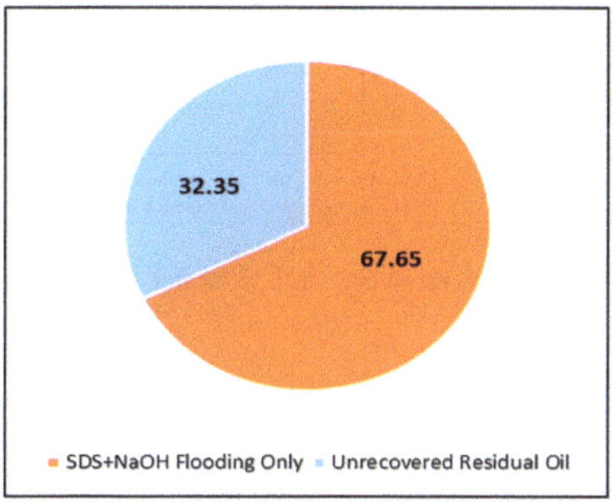

Fig. 10. Residual oil recovery (%) from M#1 core by F2 CEOR technique

Figure 8 show results from F1 CEOR technique, it can be studied that flooding of SDS recovered about 32.05% of residual oil volume. Then, very less percentage of 1.28% of residual oil was recovered by flooding of NaOH. Besides, 23.08% of residual oil was recovered by flooding of SDS + NaOH slug. In total, flooding by F1 CEOR technique recovered about 56.41% of residual oil. Hence, F1 CEOR flooding technique was found to be less efficient in higher percentage recovery of residual oil as exactly 43.59% of residual oil was left in the porous media of the reservoir being unrecovered.

Meanwhile, results from F2 CEOR technique in Fig. 10, it can be studied that flooding of only SDS + NaOH slug recovered about 67.65% of residual oil volume. Hence, F2 CEOR flooding technique was found to be more efficient in recovery higher percentage of residual oil as exactly 32.35% of residual oil was left in the porous media of the reservoir being unrecovered compared to F1 CEOR technique (Fig. 11).

Therefore, it can be deduced that flooding of only SDS + NaOH slug promises higher efficiency residual oil recovery. This may because of the S + A slug flooding in F2 CEOR technique reduce the adsorption of S into porous media of formation and promoted more IFT reduction between aqueous and oleic phase that led to higher oil recovery compared to F1 CEOR flooding. Besides, F2 CEOR process also saves cost as there is no need for flooding of SDS and NaOH as in F1 CEOR technique.

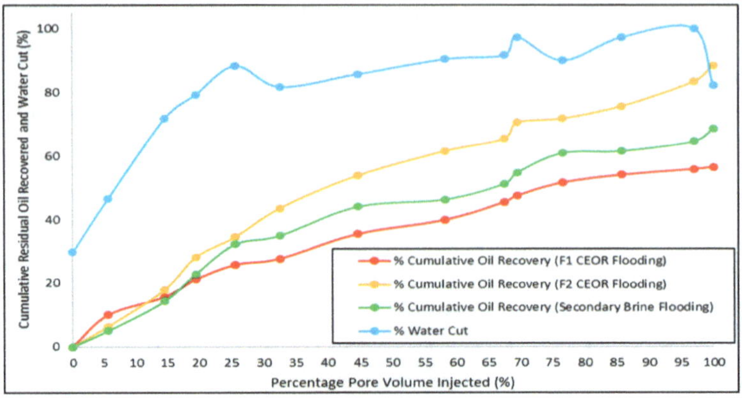

Fig. 11. Percentage cumulative oil recovery and % water cut during flooding of secondary brine, F1 and F2 CEOR techniques

4 Conclusion

The CMC of SDS in 6940 ppm of brine were found to be 1.6 wt./V% and with IFT value of 0.8 mN/m. Meanwhile, 2.0 wt./V% is the CMC for NaOH in 6940 ppm of brine as the IFT value for that concentration of NaOH showed IFT value of 0.8 mN/m. The formulated SDS + NaOH slug was a surfactant alkali slug with concentration of 1.6 wt./V% of SDS and 2.0 wt./V% of NaOH in 6940 ppm of brine showed the lowest IFT reading of 0.6 mN/m. This formulated slug is the suitable surfactant alkali slug to flood the Moran core plug by means of chemical enhanced oil recovery flooding techniques in order to improve EOR of depleting Moran Oil Field reservoirs. Thus, core flooding experiments were conducted at reservoir temperature of 78 °C by means of two CEOR flooding techniques which are referred as Flooding 1 CEOR and Flooding 2 CEOR techniques. Flooding 1 CEOR technique is the flooding of SDS, then NaOH and followed with SDS + NaOH slug. Meanwhile, Flooding 2 is the CEOR technique where the Moran sandstone core is flooded with only SDS + NaOH slug. The irreducible water saturation for the core sample from Moran Oil Field was determined to be 17.56%. Whereas residual oil saturation was perused to be 77.05%. The sandstone core plug obtained from Moran Oil Field was first flooded with 1.6 wt./V% of 0.5 Molar of SDS, then flooded with 2.0 wt./V% of 1.0 Molar of NaOH followed by 1.6 wt./V% SDS + 2.0 wt./V% NaOH slug (Flooding 1 CEOR technique) recovered approximately 56.41% of residual oil trapped in porous media of Moran formation which cannot be recovered by means of secondary brine flooding which represents secondary recovery mechanism. Besides, the overall recovery efficiency from secondary brine flooding and Flooding 1 CEOR technique was 86.25%. Comparing with Flooding 2 CEOR technique which was flooding of only 1.6 wt./V% SDS + 2.0 wt./V% NaOH in the Moran sandstone recovered exactly 67.65% of residual oil trapped in porous media of Moran formation which cannot be recovered by means of conventional methods. Hence, the overall recovery efficiency from secondary brine flooding and Flooding 2 CEOR technique was 89.81%. therefore, it can be concluded that only 1.6 wt./V% of SDS + 2.0 wt./V% of NaOH recovered more residual oil volume and found to be more effective technique compared to flooding of 1.6 wt./V% of SDS, then 2.0 wt./V% of NaOH followed by flooding of 1.6 wt./V% of SDS + 2.0 wt./V of NaOH slug. This is because NaOH reduces the wettability of the reservoir from oil wet or less water wet reservoirs to more water wet reservoirs. Moreover, NaOH in the SDS + NaOH slug reduces the adsorption of SDS surfactant into porous media of Moran core. Thus, the SDS surfactant has higher tendency to dissolve in the aqueous phase of the reservoir rather than getting absorbed in porous media of Moran formation and reduce the interfacial tension between aqueous and oleic phase. The reduction in interfacial tension between aqueous and oleic phase will mobilise the trapped residual oil as well as increases the displacement efficiency, E_D of oil.

Acknowledgements. We would like to acknowledge the Indo-US Project F. No. 194-1/2009 (IC) dated 20.2.15 titled "Foam-assisted CO_2 Flooding for the depleted Reservoirs of Upper Assam Basin and in Candidate Reservoirs in Louisiana" of 21st Century Knowledge Initiative Program under University Grants Commission. Thanks are also extended to all individuals associated with the project.

References

Das, B.M.: Microbial enhanced oil recovery: a comprehensive approach. Everyman's Sci. **4**, 234–237 (2012)

Elias, R.: Orcutt oil field thermal diatomite case study: cyclic steam injection in the careaga lease, Santa Barbara County, California. Society of Petroleum Engineers (2013)

Hakiki, F., Maharsi, D.A., Marhaendrajana, T.: Surfactant-polymer coreflood simulation and uncertainty analysis derived from laboratory study. J. Eng. Technol. Sci. **47**(6), 706–725 (2016). https://doi.org/10.5614/j.eng.technol.sci.2015.47.6.9

Hobson, D.G.: Introduction to Petroleum Geology. Scientific Press (1975). ISBN 9780901360076

Walsh, M., Lake, L.W.: A Generalized Approach to Primary Hydrocarbon Recovery. Elsevier, Amsterdam (2003)

Smith, C.R.: Mechanics of Secondary Oil Recovery. Krieger Publishing Corporation (1996)

Palo Alto, C.A.: Enhanced Oil Recovery Scoping Study. Electric Power Research Institute. Final Report, No. TR-113836 (1999)

Pillon, L.Z.: Interfacial Properties of Petroleum Products. CRC Press, Boca Raton (2008)

Martin, F.D., Oxely, J.C., Lim, H.: Enhanced recovery of a "J" sand crude oil with a combination of surfactant and alkaline chemicals. In: 60th Annual Fall Meeting, Las Vegas, SPE 14293 (1985)

Nelson, R.E., Lawson, J.B., Thigpen, D.R., Stegemeier, G.L.: Cosurfactant-enhanced alkaline flooding. In: 4th Symposium on Enhanced Oil Recovery, Tulsa. SPE/DOE 12672 (1984)

Krumine, P.H., Falcone Jr., J.S., Campbell, T.C.: Surfactant flooding 1: the effect of alkaline additives on IFT, surfactant adsorption and recovery efficiency. SPEJ **22**, 503–513 (1982)

Krumine, P.H., Falcone Jr., J.S., Campbell, T.C.: Surfactant, polymer and alkali interactions in chemical flooding processes. In: International Symposium of Oilfield and Geothermal Chemistry, Denver, SPE 11778 (1983)

Reisberg, J., Doscher, T.M.: Interfacial phenomena in crude oil-water systems. Prod. Mon. **21**, 43–50 (1956)

Gogoi, S.B., Das, B.M.: Use of an effluent for enhanced oil recovery, pp. 367–370 (2012)

Gurgel, A., Moura, M.C.P.A., Dantas, T.N.C., Neto, E.L.B., Neto, A.A.D.: A review on chemical flooding methods applied in enhanced oil recovery. Braz. J. Petrol. Gas **2**, 83–95 (2008)

Mandal, A.: Chemical flood enhanced oil recovery: a review. Int. J. Oil Gas Coal Technol. **9**, 241–264 (2015)

Nagalajan, R., Wasan, D.C.J.: Measurement of dynamic interfacial tension by an expanding drop tensiometer. Colloid Interface Sci. **159**, 164–173 (1993)

Shen, P., Zhu, B., Li, X.B., Wu, Y.S.: An experimental study of the influence of interfacial tension on water-oil two phase relative permeability. Transp. Porous Media **85**, 505–520 (2010)

A Green Approach for Oil Field Produced Waters of Upper Assam Basin

Gogoi J. Tapan, Gogoi B. Subrata(✉), and Sarmah Shilpi

Department of Petroleum Technology, Dibrugarh University, Dibrugarh, India
tapanjyotigogoil3@gmail.com, subrata@dibru.ac.in

Abstract. The focus in oil industry is to carry out exploration and exploitation to develop hydrocarbon prospects for the commercial production of oil and natural gas in an environmentally harmonious manner. While producing liquid and gaseous hydrocarbons, large amount of associated water is also produced. In majority of the cases in Upper Assam Basin the oil wells are producing with more than 85% water cut. This produced water has to be treated before discharging it in the environment as it contains harmful constituents in the form of both organics and inorganics. In this study, five samples have been taken for analysis with water content of 94%, 92%, 90%, 84% and 82% which shows the presence of oil and grease, TDS, TSS, DO, Ca, Fe, Na, K, Li, Mg, Mn, Pb and Zn. However the presence of Cr, Cu, Mo, and Ni have not been observed in the samples.

In the present global scenario, much awareness is observed in the environment sector. People are aware of the worst consequences of the environment hazards. Several factors leads to such consequences, industrial waste water being one of them. The direct discharge of the waste water from the industries causes harmful impact on flora and fauna. Therefore, this paper primarily focuses on the treatment processes of the industrial waste water and suggests some novel green approaches to be implemented by the industries for the safe disposal. Hence the samples under study have been treated with microfiltration, ultrafiltration and nanofiltration in the Hollow Fibre Membrane Module set up. The different test samples were collected from Oil fields of Upper Assam Basin. When the samples were treated with microfiltration it was found that some of the parameters like TDS, Turbidity, D.O, Salinity, Oil & Grease and some heavy metals were not within the limits set by the pollution control board of India. Therefore, the samples were further treated with Ultrafiltration. This treatment process resulted in the treatment of most of the pollutants. However turbidity, D.O., Salinity, Oil & Grease were still not within the range. The third treatment process of nanofiltration resulted in the complete eradication of pollutants and all the parameters were found to be within the specified range set by the pollution control board of India. Hence after completion of the three stages of treatment processes consecutively, the treated water may be safely disposed of without any harmful consequences.

Keywords: Produced water · Pollutants · Organics · Inorganic

H. Ameen and T. Sorour (Eds.): GeoMEast 2018, SUCI, pp. 151–171, 2019.
https://doi.org/10.1007/978-3-030-01929-7_11

Abbreviation

AAS – Atomic Absorption Spectrometer
BOD – Biochemical Oxygen Demand
CO – Crude Oil
COD – Chemical Oxygen Demand
CPCB – Central Pollution Control Board
DO – Dissolved Oxygen
EOR – Enhanced Oil Recovery
FP – Flame Photometer
ICMR – Indian Council of Medical Research
ICP-OES – Induced Coupled Plasma Optical Emission Spectrophotometer
MINAS – Minimum National Standard
MF – Microfiltration
MWCO – Molecular Weight Cut-off
NF – Nanofiltration
O&G – Oil and Grease
PW – Produced Water
TDS – Total Dissolved Solid
TSS – Total Suspended Solid
UF – Ultrafiltration
US – United States
UV – Ultra Violet
WHO – World Health Organization

1 Introduction

Water is generally present with oil in the reservoir. As the oil production continues its saturation slowly decreases and the water saturation increases which in later stage is produced with oil sometimes in the form of emulsion. This water besides containing hydrocarbons in soluble, emulsified and free form, also usually is high in suspended and dissolved solids. Hydrocarbons are considered among the most harmful pollutants as they are lethal to human, animal and plant on direct consumption.

It is thus, obvious that the effluent cannot be let off as such in the environment. Realizing the threat posed by industrialization and rapid development, government all over the world have imposed strict legislations to the industries. Various agencies such as United States (US) environmental protection agency, World Health Organisation (WHO), Central Pollution Control Board (CPCB) of India, Indian Council of Medical Research (ICMR) and Minimum National Standard (MINAS) have formulated standards for quality of water used for different purpose and for safe disposal by various means.

From natural gas production operations, Produced Water (PW) account for 80% of the wastes and residuals produced. Besides oil wells are producing with 98% PW and 2% fossil fuel in some oil fields of Upper Assam Basin [1]. For each barrel of oil, 7 to 10 barrels of PW is produced in US [2]. After treatment the PW can be discharged into the environment or can be used in water injection wells for reservoir pressure maintenance [2, 3]. The PW quality vary considerably depending on the location of the reservoir oil field, the characteristic and type of the producing well, the geochemistry of the producing formation and the type of hydrocarbon product being produced [2–5].

The membrane separation is a technology which separates two fluids via pores or gaps in the molecular arrangement of a continuous structure. This technology is being used for treatment of contaminated water so as to bring the level of contamination within the limits set by the pollution control board of both the Centre and state of India.

2 Experimental

2.1 Materials

Five samples of formation water were collected from oil fields of Upper Assam Basin (Table 1).

Table 1. Materials used

Sl. No.	Materials	Specification	Firm
1	Formation Water 1	Barail formation	Oil Industry
2	Formation Water 2		
3	Formation Water 3		
4	Formation Water 4		
5	Formation Water 5		
6	Potassium hydroxide (KOH)	M.W. – 56.11 g/mol	Merck specialities private limited
7	Ethanol (C_2H_5OH)	M.W. – 46.07	Biochem life sciences
8	Petroleum Ether	Wt. Per ml at 20 °C, 0.630–0.645 g	Fisher Scientific
9	Ethylene di-amine tetra acetic acid (EDTA) Disodium salt dihydrate	0.01 M	RFCL Limited
10	Erichrome Black T	0.5 g in 100 ml	RFCL Limited
11	Potassium Chloride (KCl)	M.W. – 74.56 g/mol	Avantor Performance Materials India Limited
12	Sodium Chloride (NaCl)	M.W. – 58.44 g/mol	RFCL Limited
13	Calcium Carbonate ($CaCO_3$)	M.W. – 100.09 g/mol	Merck specialities private limited
14	Lithium Carbonate (Li_2CO_3)	M.W. – 73.89 g/mol	Spectrochem Pt. Ltd. Mumbai (India)

(*continued*)

Table 1. (*continued*)

Sl. No.	Materials	Specification	Firm
15	Hydrochloric Acid (HCl)	M.W. 36.46 g/mol, Sp. Gr. at 25 °C is about 1.18	RFCL Limited
16	Phenolphthalein	pH range 8.2–10 (colourless to pink)	RFCL Limited
17	Filter Paper	Pore size 0.45 μm, diameter 47 mm	Whatman
18	Toluene (C_7H_8)	M. W. – 92.14 g/mol	Merck Specialities Private Limited

2.2 Methods

The following instruments were used for the experimental work (Table 2).

Table 2. Parameters determined

Sl. No.	Parameters	Instruments	Chemicals used
1	pH	Make - Systronics Model - Water Analyser371	–
2	Salinity		
3	Conductivity		
4	Total Dissolved Solids (TDS)		
5	Turbidity		
6	Dissolved Oxygen (DO)		
7	Total Suspended Solids (TSS)	Make - Vacuubrand Model - D 97877	–
8	Water Content	Make - Electra Model - Electra Lab Centrifuge	Toluene (C_7H_8)
9	Biochemical Oxygen Demand	Make - VELP Scientifica Model - FOC	Potassium Hydroxide (KOH)
10	Sodium (Na)	Make - Systronics Model - Flame Photometer128 Compressor 126	Sodium Chloride (NaCl)
11	Potassium (K)		Potassium Chloride (KCl)
12	Calcium (Ca)		Calcium Carbonate ($CaCO_3$)
13	Lithium (Li)		Lithium Carbonate (Li_2CO_3)

(*continued*)

Table 2. (*continued*)

Sl. No.	Parameters	Instruments	Chemicals used
14	Manganese (Mn)	Atomic Absorption Spectrometer Make - Perkin Elmer Model - Analyst 200	–
15	Iron (Fe)		
16	Chromium (Cr)		
17	Copper (Cu)		
18	Magnesium (Mg)	Inductively Coupled Plasma Optical Emission Spectrophotometer Make - Perkin Elmer, USA Model - Optima 2100 DV	–
19	Molybdenum (Mo)		
20	Nickel (Ni)		
21	Lead (Pb)		
22	Strontium (Sr)		
23	Zinc (Zn)		
24	Membrane separation	Hollow Fiber Membrane Module Set Up Make - IIT Kharagpur Model - HFMS – 01	–

3 Experiments

The following experiments will be performed both before and after treatment.

3.1 Water Analyzer

The pH, conductivity, salinity, TDS, Dissolved Oxygen (DO) and turbidity by using Systronics water analyser 371.

3.2 Water Content

For evaluating water content, the oldest and most widely used method is the centrifuge test, where equal volumes of oil and solvent are placed in a graduated centrifuge tube. After centrifugation, the volume of the higher gravity water and sediment layer at the bottom of the tube is read. Therefore, 5 ml of CO + 5 ml of toluene is completely mixed. The new sample is run in the centrifuge for 5 min. And then two layers is easily visible. It can be calculated by using the formula

$$\text{Water content} = \frac{\textit{Amount of water sample visible}}{\textit{Amount of sample initially taken}} * 100\%. \tag{1}$$

3.3 Oil & Grease (O&G)

20 ml sample water + 40 ml petroleum ether in separating funnel was mixed and shaken for a half an hour. Appearance of two layers was observed where the lower

layer was separated out in the beaker then dried in the oven and weighed = y, x = weight of the empty beaker.

$$\text{O\&G (ppm)} = \frac{(y - x) \times 10^6}{sample} \tag{2}$$

3.4 Total Suspended Solid (TSS)

The TSS can be calculated by using Millipore assembly. It can be calculated by using the formula below

$$\text{TSS (ppm)} = \frac{Wt\ of\ filter\ paper\ with\ cake - Wt\ of\ filter\ paper}{Volume\ of\ water\ filtered\ in\ 30\ mins}. \tag{3}$$

3.5 Biochemical Oxygen Demand (BOD)

The BOD can be calculated by measuring the dissolved oxygen (DO) before putting the sample in the BOD incubator for 5 days. After 5 days DO is calculatedagain. The difference gives the value of BOD.

3.6 Titration Methods

Alkalinity: 50 ml sample water + methyl orange titrated with 0.05 N HCl = T

$$HCO_3 = \frac{T \times 0.05 \times 50 \times 1000}{sample} \tag{4}$$

$$\text{Total Alkalinity (ppm)} = HCO_3 \times 0.82 \tag{5}$$

Total hardness: 50 ml sample water + buffer solution (3–4 drops) + Erichrome black indicator, then titrated with EDTA 0.01 M or 0.02 N. A change to ink colour was observed after titration.

$$\text{Total hardness (ppm)} = \frac{EDTA \times 1000}{sample} \tag{6}$$

3.7 Flame Photometer (FP)

Flame Photometer instrument is used to determine Na, K, Li and Ca present in the FW.

Preparation of stock standard solution

a. Na: A standard solution of 1000 ppm is prepared by dissolving 2.5416 g NaCl in one litre of distilled water.
b. K: A standard solution of 1000 ppm is prepared by dissolving 1.9070 g KCl or 2.5869 g KNO_3 in one litre of distilled water.

c. Ca: A standard solution of 1000 ppm is prepared by dissolving 2.497 g $CaCO_3$ in approx 300 ml glass distilled water and adding 10 ml conc. HCl in one litre of distilled water.
d. Li: A standard solution of 2000 ppm is prepared by dissolving 4.945 g Li_2CO_3 in approx 300 ml glass distilled water and adding 15 ml conc. HCl in one litre of distilled water.

3.8 Atomic Absorption Spectrometer (AAS)

Determination of Fe, Cu, Mn and Cr of PW was conducted in the instrument Atomic Absorption Spectrophotometer (AAS). Prior to the determination of the amount present, the AAS equipment was calibrated and analyzed using blank solution of 2% Nitric acid (HNO_3) solution and standard solutions of 1 ppm, 2 ppm, 3 ppm and 4 ppm of Fe solution for Fe test. Similarly for Cr, Mn and Cu 1 ppm, 2 ppm, 3 ppm and 4 ppm of Cr, Mn and Cu standards samples were made respectively. After calibration samples were aspirated, the results were obtained.

3.9 Inductively Coupled Plasma Optical Emission Spectrometry (ICP-OES)

In ICP-OES when plasma energy is given to an analysis sample from outside, the component elements are excited. When the excited atoms return to low energy position, emission rays are released and the emission rays that correspond to the photon wavelength are measured. The element type is determined based on the position of the photon rays, and the content of each element is determined based on the rays' intensity. The elements such as Mg, Mo, Ni, Pb, Sr and Zn of PW were determined using this instrument.

3.10 Hollow Fiber Membrane Module Set Up

The Hollow Fiber Membrane Module Set Up mainly consists of the hollow fiber module. The feed is drawn by the booster pump and fed to the module by 6 mm polyurethane tube via a Perspex flange. Two pressure gauges in the range of 0 to 60 psi are attached to the upstream and downstream of the module. A ¾" needle valve of stainless steel has been fitted in the retentate line after the module. This valve is used for fine tuning of pressure and flow rate through the module. A rotameter of range 0 to 50 l/h is attached to the retentate line and the retentate stream is recycled back to the feed tank. A bypass line is connected from the pump to the feed tank and a 1/2" stainless steel needle valve is attached to the bypass line. The permeate flows through a 5 mm polyurethane pipe into permeate collector.

4 Results

4.1 Results of Water Content

The oil fields of Upper Assam Basin produces crude oil with high water cuts of more than 80% [6]. From the above results we can easily evaluate that the oil fields of Upper Assam Basin are producing a huge amount of water along with crude oil (Table 3).

Table 3. Values of water content

Sl. No.	Sample ID	Geologic formation/depth (m)	Reservoir pressure (kg/cm^2)	Reservoir temperature (°C)	Water cut i.e. water in crude oil (%)
1	Sample 1	Barail (3200)	240	87	90
2	Sample 2	Barail (3100)	200	88	82
3	Sample 3	Barail (3200)	240	87	94
4	Sample 4	Barail (3250)	190	85	92
5	Sample 5	Barail (3100)	200	88	84

As the water produced with oil cannot be directly disposed in the environment, so these water has to be treated properly to meet the standards imposed by the pollution control boards of both the Centre and state. The physical, chemical and biological parameters were evaluated and gradually the PW has been treated in the Hollow Fiber Membrane Module set up for Microfiltration (MF), Ultrafiltration (UF) and Nanofiltration (NF).

MF, UF and NF utilize high pressure across the membranes to accomplish filtration of contaminants from the PW. These technologies are the most common techniques of water purification. MF, UF and NF are based on the principle of rejection of species higher than the pore size of the membrane under pressure.

4.2 Results of Physical and Chemical Analysis

4.2.1 Results of Water Analyzer

The chemical and biological properties of PW are affected due to a high value of pH where we can see the alkaline nature in above cases [6]. But after treatment the values proceeds towards normal. The carbon dioxide/bio-carbonate/carbonate governs the pH of PW [6]. The conductivity of PW is well above that of pure water whose conductivity is 0.35 ppm due to the presence of ions in it [6]. One of the most important challenge of water treatment system is the removal of dissolved solid, salts or impurities. TDS gives the sum of all ion particles which are smaller than 2 micron and ranges from <2000 ppm to >150,000 ppm [8]. But after treatment in the hollow fiber membrane by ultrafiltration and nanofiltration the values are less than 2100 ppm which is below the range set by the pollution control board of India. The TDS accounts for organic solutes such as hydrocarbons and sodium, calcium, chloride and bicarbonates [8]. The partial removal of TDS can be done by ion exchange and precipitation methods but for complete removal of all dissolved solids from water can be done through filtration processes like reverse osmosis. However, this process is not intended to be used for wastewater treatment, but to provide drinking water from seawater in desalinization plants as of its high costs (Figs. 1, 2, 3, 4, 5 and 6).

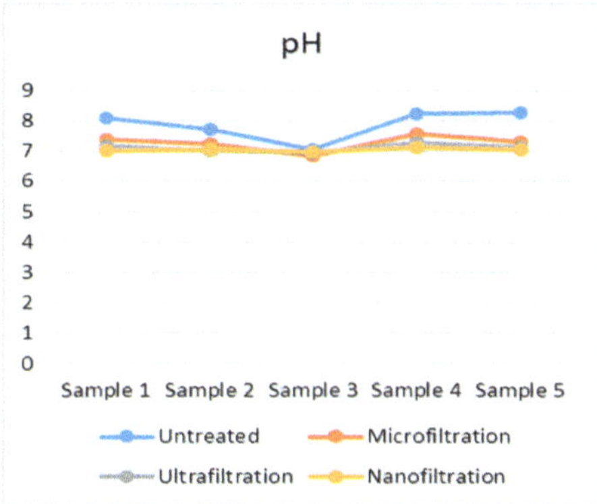

Fig. 1. Comparison of pH after stages of membrane treatment

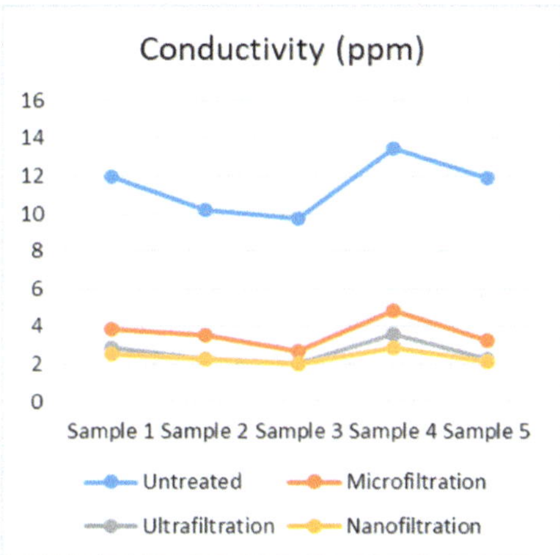

Fig. 2. Comparison of Conductivity after stages of membrane treatment

Conductivity is a process to measure the capacity of water to pass electrical flow which is related to the concentration of ions in the water. The conductive ions basically come from dissolved salts and inorganic materials such as alkalis, chlorides, sulphides and carbonate compounds. With the higher presence of ions the conductivity of ions becomes higher which is being measured at 25 °C. The turbidity measurements are useful to determine the optimum amount of coagulants to treat industrial wastewaters. Also it is used to evaluate the performance of water treatment plants.

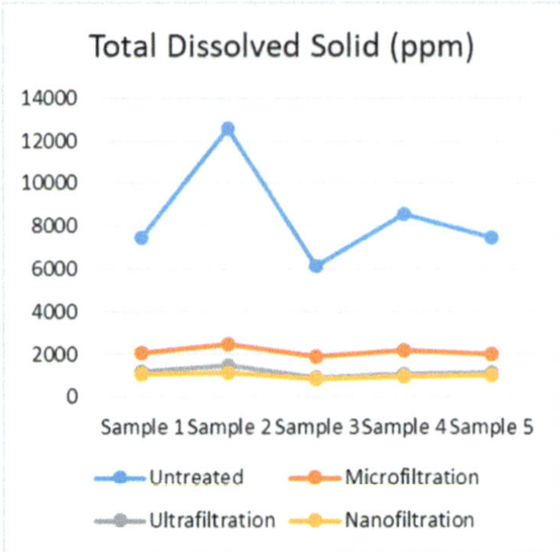

Fig. 3. Comparison of Total Dissolved Solid after stages of membrane treatment

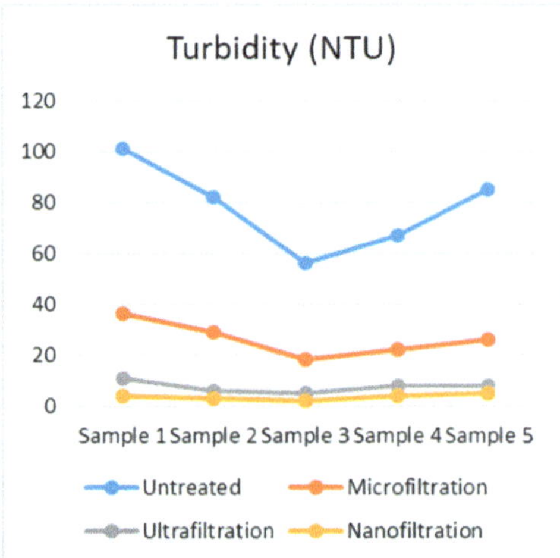

Fig. 4. Comparison of Turbidity after stages of membrane treatment

The PW is highly saline which is very rich in dissolved minerals [6]. Salinity is the total concentration of all dissolved salts in water whose range is from 7000–9000 ppm. But after treatment the values comes well below 1000 ppm. The major contributing ions for salinity are sodium, magnesium, calcium, potassium, chloride, sulphate, bicarbonate and bromine. DO analysis measures the amount of gaseous oxygen dissolved in an aqueous solution. The diffusion process leads to entry of oxygen into the water from the surrounding air. Higher level of total dissolved solid gas concentration can be harmful to aquatic life [10]. But, adequate DO is necessary for good water quality.

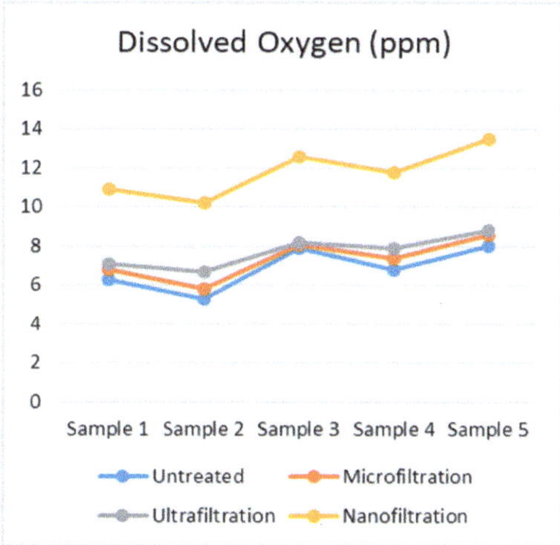

Fig. 5. Comparison of Dissolved Oxygen after stages of membrane treatment

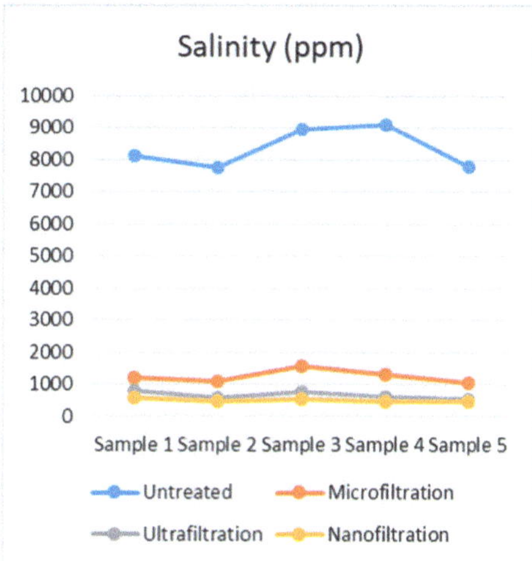

Fig. 6. Comparison of Salinity after stages of membrane treatment

4.2.2 Results of Millipore Assembly

The concentration and composition of suspended solids are fundamental indicators of water quality. In a water flood process possibility exists that these suspended solids will cause the injection wells to become impaired. Therefore, selection of specific process or combined processes for removal of suspended solids from water depends on the character of the solids, their concentration and the required filtrate clarity. The main physical processes of formation damage in water injection wells can result from plugging by solids present in injection water (Fig. 7).

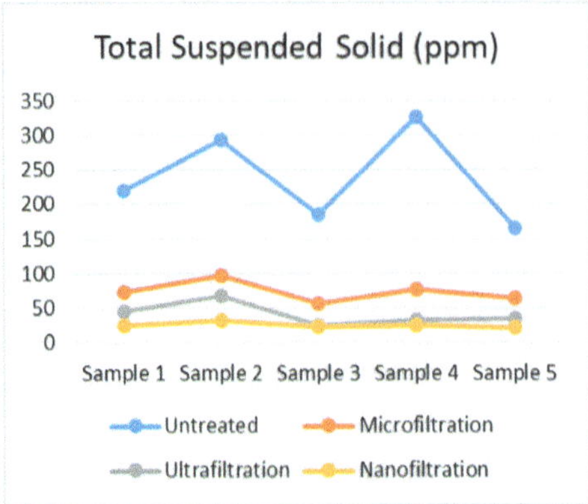

Fig. 7. Comparison of Total Suspended Solids after stages of membrane treatment

4.2.3 Results of Titration Methods

The sum of total cation present in water give us the hardness which is perhaps due to minerals coming from underground reservoir and drilling activities [6]. It is predominantly caused by divalent cation such as Calcium, Magnesium and others (e.g. aluminum, barium, iron, manganese, strontium and zinc). The dissolved polyvalent metallic ions from sedimentary rocks, seepage and runoff from soils are the principal sources of hardness in water [9]. Alkalinity is primarily due to hydroxide, carbonate and bicarbonate contents (Figs. 8 and 9).

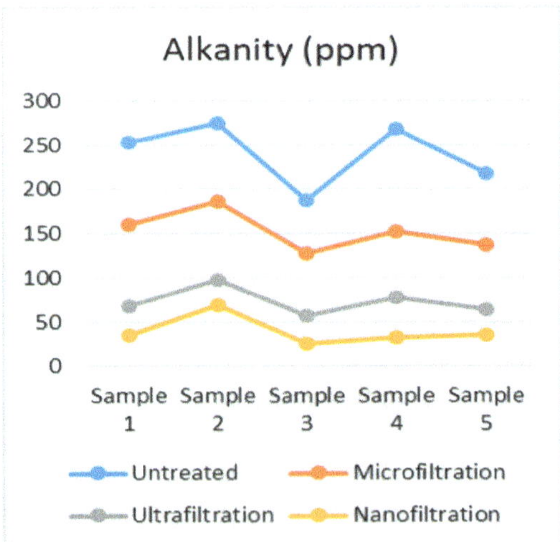

Fig. 8. Comparison of Alkalinity after stages of membrane treatment

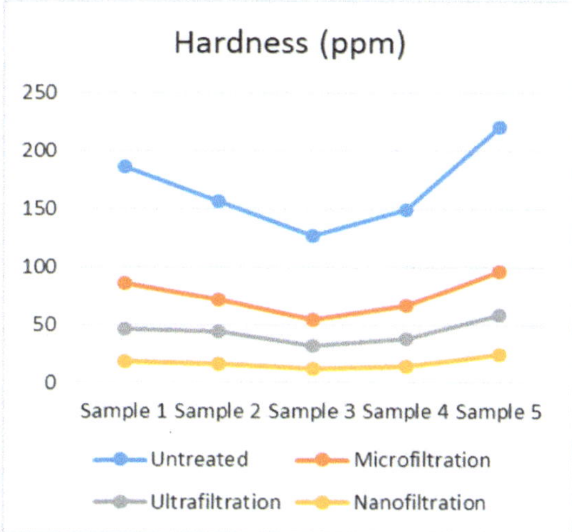

Fig. 9. Comparison of Hardness after stages of membrane treatment

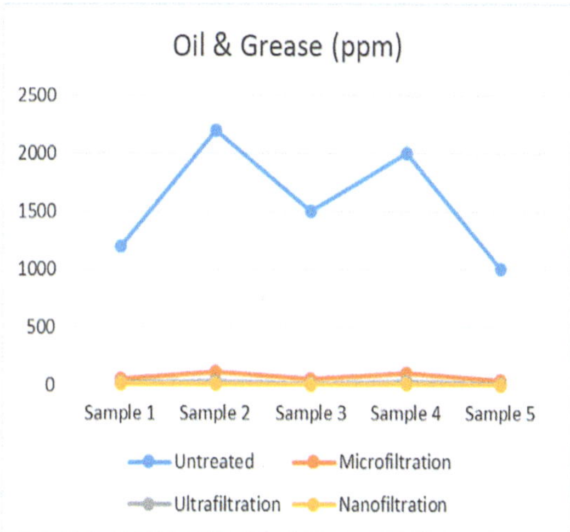

Fig. 10. Comparison of Oil & Grease after stages of membrane treatment

4.2.4 Results of Oil and Grease (O&G)

The hydrocarbons as oil and grease is present in PW in the form of free oil, dispersed oil and emulsified oil. The disposal limit of O&G is 10 ppm but the values of raw PW is very high, even after treatment by MF and UF all the values are not within range. But after NF all the values were within range. The separation of two immiscible liquids is governed by Stokes law which gives the rate of fall of a small sphere through a viscous fluid. Stokes law states that when this sphere is under the influence of gravity it attains a constant velocity, which is given by the stokes equation (Fig. 10)

$$Vt = \frac{\{(\rho - \rho o)gdo2\}}{18\mu} \quad (7)$$

Where,

V_t – Terminal velocity of the drop
ρ – Water density
ρ_o – Oil density
g – Gravitational acceleration
d_o – oil drop diameter
μ – water viscosity

4.3 Results of Biological Analysis

4.3.1 Results from BOD Incubator

It is a measure of the quantity of oxygen used by microorganisms in the oxidation of organic matter. When there are high levels of contaminants in PW then the microorganisms consume more oxygen to decompose the organic matter, and hence it decreases the amount of oxygen required by other aquatic organisms to live (Fig. 11).

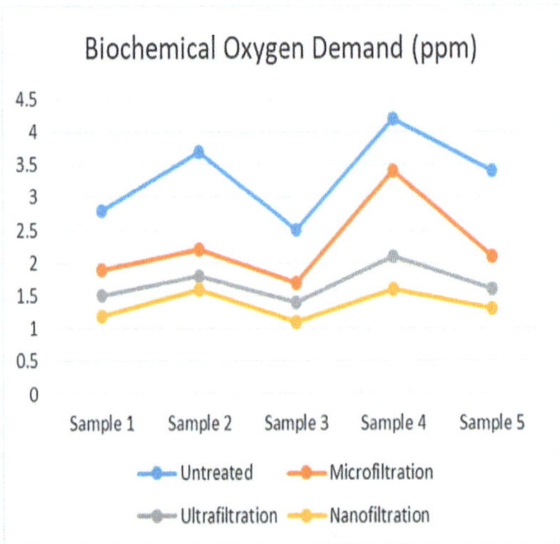

Fig. 11. Comparison of Biochemical Oxygen Demand after stages of membrane treatment

4.4 Results of Inorganics

4.4.1 Results of Flame Photometer (FP)

High levels of salts present in PW adversely affect organic removal efficiency and sludge settle ability during the treatment process. Also it cannot be disposed in the environment, which will increase the salinity of the water where it is being disposed [11]. The elements Na, K and Ca are naturally occurring elements which is present in

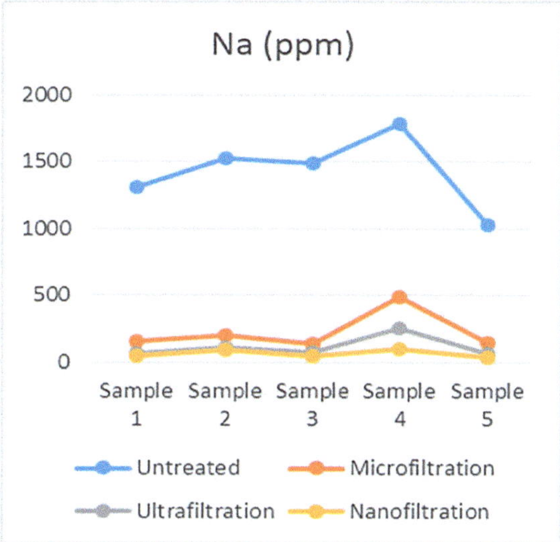

Fig. 12. Comparison of Sodium (Na) after stages of membrane treatment

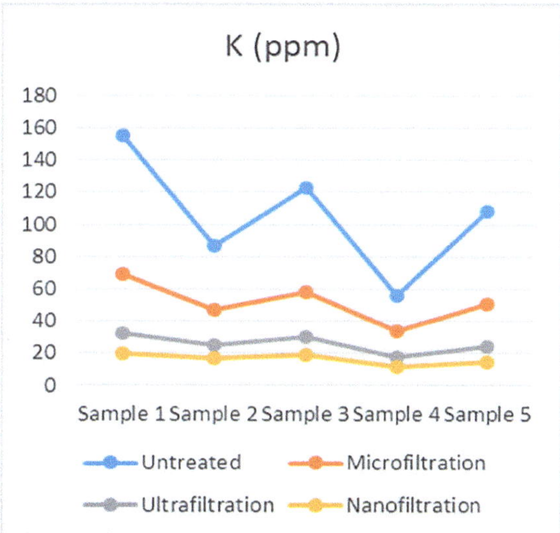

Fig. 13. Comparison of Potassium (K) after stages of membrane treatment

natural water from the weathering of various rocks. But these quantities increase in PW due to contamination of hydrocarbons. Therefore it has to be treated to bring down the parameters within its permissible limits (Figs. 12, 13, 14 and 15).

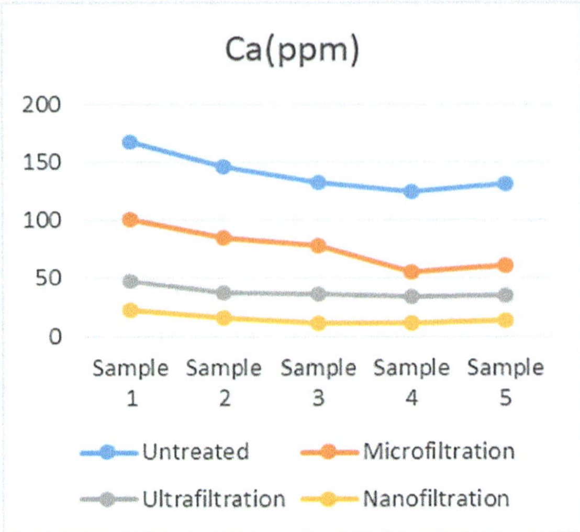

Fig. 14. Comparison of Calcium (Ca) after stages of membrane treatment

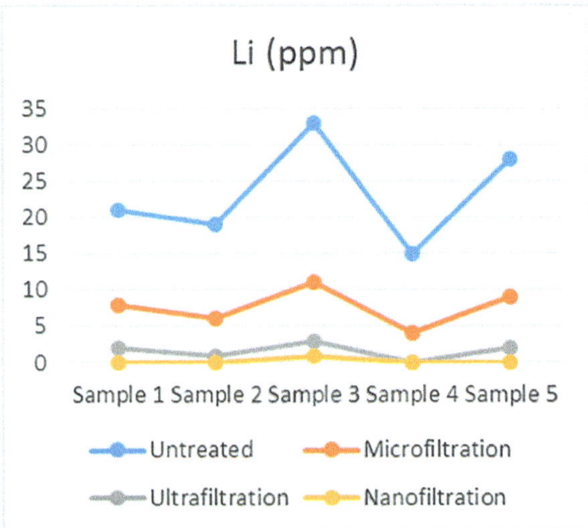

Fig. 15. Comparison of Lithium (Ca) after stages of membrane treatment

4.4.2 Results of Atomic Absorption Spectrometer (AAS)

Natural water contains variable amounts of iron but in ground water it is normally present in the ferrous or bivalent form (Fe^{++}). The quality varies depending upon the geological area and it is required both by plants and animals [10]. High levels of Fe and Mn in PW can promote the growth of bacteria which obtains its energy from the

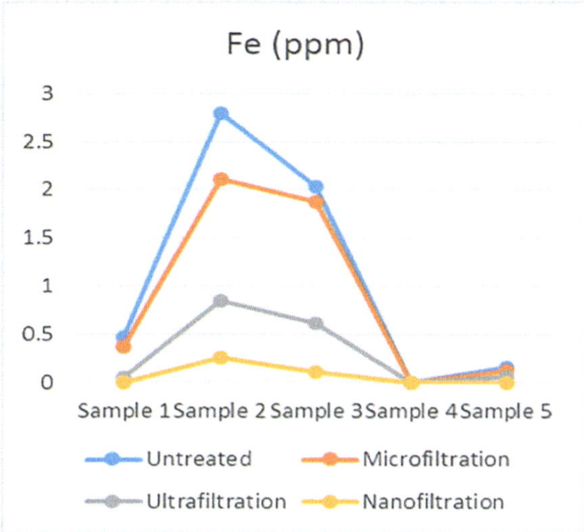

Fig. 16. Comparison of Iron (Fe) after stages of membrane treatment

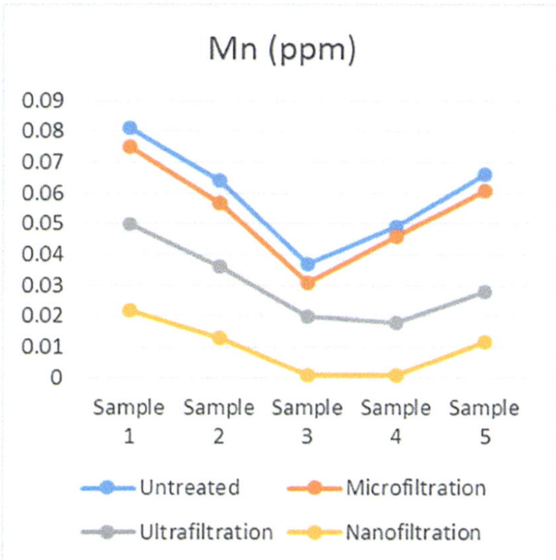

Fig. 17. Comparison of Manganese (Mn) after stages of membrane treatment

chemical reaction that occurs when Fe and Mn mixed with DO. It form thick slime growths on the walls of the pipe which finally mixes with the PW. The reduced levels of Fe after treatment determines that the Fe is fully oxidized (Figs. 16 and 17).

4.4.3 Results of Inductively Coupled Plasma Optical Emission Spectrophotometer (ICP-OES)

The heavy metals are directly related with environmental pollution and biological toxicity problems [12]. The metallic elements are human carcinogens which may induced multiple organ damage even at lower levels of exposure [13] (Figs. 18, 19, 20 and 21).

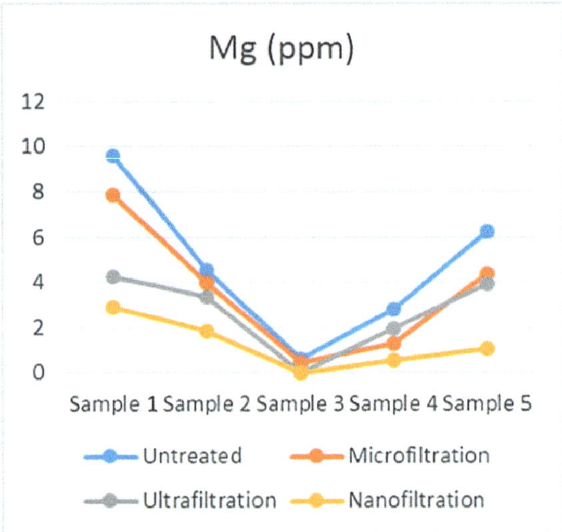

Fig. 18. Comparison of Magnesium (Mg) after stages of membrane treatment

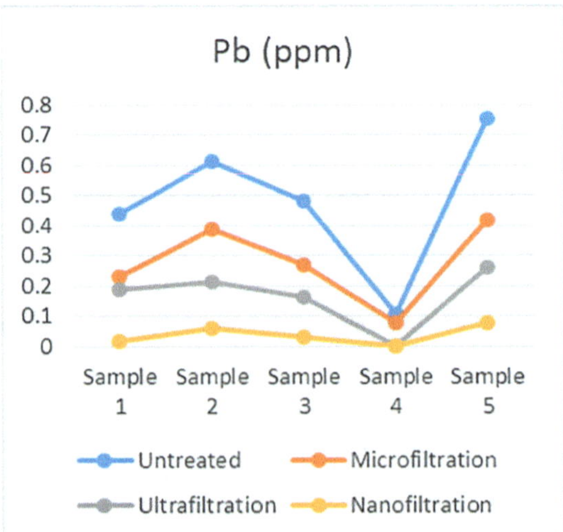

Fig. 19. Comparison of Lead (Pb) after stages of membrane treatment

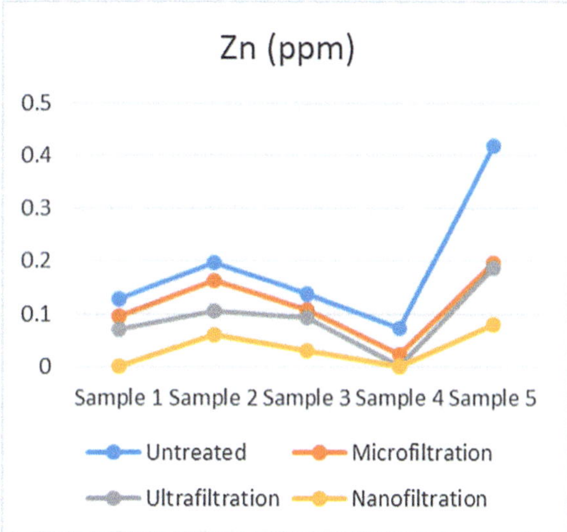

Fig. 20. Comparison of Zinc (Zn) after stages of membrane treatment

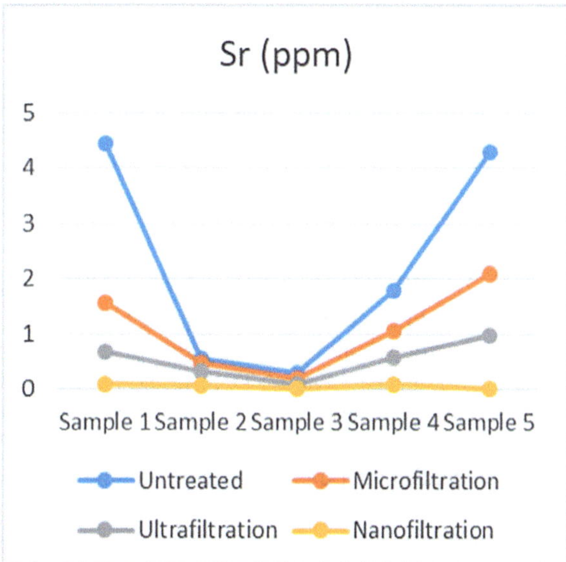

Fig. 21. Comparison of Strontium (Sr) after stages of membrane treatment

5 Conclusion

In any case if the effluent is discharged into environment without treatment, it will cause pollution, affecting the salinity, BOD, COD and many more of the water sources into which they may be discharged on contaminated land. The effluents reduce DO concentration and formation of sludge deposits leading to environmental disturbances that may damage aquatic biota. Also injection of effluent into formation may decrease its injectivity due to plugging and choking of pores if suspended solids and oil are not removed before injection. Also improved EOR techniques are being used extensively involving polymers, surfactants and various gases that will adversely affect the interfacial properties of the produced waste waters making the resolution of emulsions more difficult in future.

In some cases, significant treatment of produced water is required to meet the quality required for beneficial uses such as irrigation, farmland restoration, cattle and animal consumption, and drinking water for private use or in public water systems [7].

To prevent scaling and water contamination from PW removal of bacteria, viruses, microorganisms, algae, etc. is very necessary. Microorganisms occur naturally in the PW or may be added during de-oiling treatments and to remove it advanced filtration techniques are one of the effective technologies. UV light treatment, chorine or iodine reaction, ozone treatment and pH reduction are other treatments available to disinfect produced water [8] (Table 4).

Table 4. Membranes with average pore diameters

Sl. No.	Membrane	Molecular weight cut-off (MWCO) (kDa)	Average pore diameter
1.	MF	–	0.2 μm
2.	UF	44	12.9 nm
3.	NF	400	1.04 nm

Treatment was done by MF, UF and NF membranes for 5 nos of PW samples collected from different oil fields of Upper Assam Basin. The average pore size of the membranes remain uniform throughout the whole membrane. While treating with MF although a significant decline in the value of TDS, turbidity, DO, salinity, O&G and heavy metals were observed, they were not within the range of CPCB (India). Again, while treating with UF the parameters such as turbidity, DO, salinity, O&G, Na, K, Pb and Sr were not within range. But after treating with NF, all the parameters were within range set by CPCB (India). The PW after treatment can be used for reservoir pressure maintenance during secondary recovery of crude oil.

Acknowledgement. We would like to acknowledge

- Indo-Tunisia Project no. DST/INT/TUNISIA/P-02/2017 entitled "Characterisation of oil field water to mitigate a major environmental hazard with recovery of usable water".
- Indo-US Project F. No. 194-1/2009(IC) dated February, 2015 entitled "Foam-assisted CO_2 Flooding for the depleted Reservoirs of Upper Assam Basin and in Candidate Reservoirs in

Louisiana" of 21st century knowledge initiative program under University Grants Commission.

- DBT Twinning Project No. BT/485/NE/TBP/2013 entitled "Remediation of produced water from NE oil field by microbial and membrane intervention with modeling of an effluent unit for the recovery of usable water".
- Dibrugarh University for the laboratory facilities

References

1. Neff, M.J., Lee, K., DeBlois, M.E.: Produced Water: Overview of Composition, Fates and Effects, January 2011
2. Guerra, K., Dahm, K., Dundorf, S.: Oil and gas produced water management and beneficial use in the Western United States, U.S. Department of the Interior Bureau of Reclamation, Science and Technology Program Report No. 157, Technical Service Center, September 2011
3. Veil, J.A., Pruder, M.G., Elcock, D., Redweik, R.J.: A White Paper Describing Produced Water from Production of Crude Oil, Natural Gas, and Coal Bed Methane Prepared by Argonne National Laboratory for U.S. Department of Energy. National Energy Technology Laboratory, January 2004
4. Joel, O.F., Amajuoyi, C.A., Nwokoye, C.U.: Characterization of formation water constituents and the effect of fresh water dilution from land rig location of the Niger Delta, Nigeria. J. Appl. Sci. Environ. Manag. **14**(2), 37–41 (2010)
5. Jacobs, R.P.W.M., Grant, E., Kwant, J., Marqueine, J.M., Mentzer, E.: The composition of produced water from shell operated oil and gas production in the North Sea. In: Ray, J.P., Englehart, F.R. (eds.) Produced Water, Technological/Environmental Issues and Solutions, pp. 13–21. Plenum Press, New York (1992)
6. Gogoi, B.S., Sen, R., Rajbongshi, A., Hazarika, K.: Characterization of oil field produced waters of upper Assam basin, India. Int. J. New Technol. Sci. Eng. **2**(1) (2015). ISSN 2349-0780
7. Lee, R., Seright, R., Hightower, M., Sattler, A., Cather, M., McPherson, B., Wrotenbery, L., Martin, D., Whitworth, M.: Strategies for produced water handling in New Mexico. Paper presented at the 2002 Ground Water Protection Council (2002)
8. Arthur, J.D., Langhus, G.B., Patel, C.: Technical Summary of Oil & Gas Produced Water Treatment Technologies, March 2005
9. Hardness in Drinking Water. Background Document for Development of WHO Guidelines for Drinking Water Quality
10. Kumar, M., Puri, A.: A review of permissible limits of drinking water. Indian J. Occup. Environ. Med. **16**(1), 40 (2012)
11. Hashad, F.M., Sharma, S., Nies, L.F., Alleman, E.J.: Study of Salt Wash Water Toxicity on Wastewater Treatment, August 2006
12. Gakwisiri, C., Raut, N., Al-Saadi, A., Al-Aisri, S., Al-Ajmi, A.: A crictical review of removal of zinc from wastewater. In: 2012 Proceedings of the World Congress on Engineering, vol. I, July 2012
13. Tchounwou, P.B., Yedjou, C.G., Patlolla, A.K., Sutton, D.J.: Heavy metals toxicity and the environment, August 2012

Potential of Using Tire Waste in Soil Reinforcement and Its Application in Geocells Under Static and Dynamic Loading – A Review

C. Rajesh Kumar and Sreevalsa Kolathayar(✉)

Department of Civil Engineering, Amrita School of Engineering Coimbatore, Amrita Vishwa Vidyapeetham, Coimbatore, India
sreevalsakolathayar@gmail.com

Abstract. The geocell retention system is one of the recent advancement techniques in geotechnical engineering. Geocells impart strength to the backfill soil by confinement, interlocking, and friction. The soil or gravel will be confined in the geocells and this confinement gives additional shear strength by its apparent cohesion. The term "geocell retaining wall" represents the stacking of the various geocell layers to form an earth retention system. The role of reinforcing layers is to restrain the earthquake-induced shear deformation in the soil. Reinforcing geocell layers provide increased stability against dynamic loading besides simply stabilizing the fill. The reinforcing geogrid layers and geocell layers resist rotational and translational failure. This paper contains brief inferences on the application of different geocells with different infill material in retention systems and retaining walls with different backfill materials. A brief overview of the usage of different types of tire waste as backfill material is presented.

1 Introduction

The fast progress of infrastructure demands effective methods of reinforcement to stabilize the soil mass. Geocells are three dimensional cellular confinement systems that provide lateral confinement to the soil. The Geocell needs to be filled with infill material and materials like sand or gravel have been put in use for the same till date. This study explores the potential of tire waste as infill materials in geocells. A review of literature on potential application of tire waste in soil reinforcement is presented in the paper along with recommendations to use them as infill materials in geocells.

1.1 Influence of Tire Chips on Soil Characteristics

Edinçliler and Ayhan (2010) stated that high compaction and deformation problems can be faced if shredded rubber tires alone are used as fill material. The addition of tire waste to sand improves its shear strength. The longer fibers have greater contact areas with the soil particles. As the aspect ratio of tire fibers increases, the shear strength of the mixture increases.

H. Ameen and T. Sorour (Eds.): GeoMEast 2018, SUCI, pp. 172–181, 2019.
https://doi.org/10.1007/978-3-030-01929-7_12

Feng and Sutter (2000) stated that adding soil to shredded rubber tires and using the combined material as backfill will turn the demerits of shredded rubber to merits, such as an increased stiffness of backfill and a reduced self-heating problem. The retaining walls with sand-shredded tire mixture as backfill will have a high capacity to absorb energy and a decreased transfer of loading energy into the trench backfill. Ni et al. (2018), Meguid and Youssef (2018) referred that the inclusion of tire-derived aggregate (Fig. 1) around the buried pipe lines will reduce the earth pressure on the pipes. The thickness of the tire-derived zone will not affect the earth pressure reduction, but the width of this zone shows a major difference on these pipes.

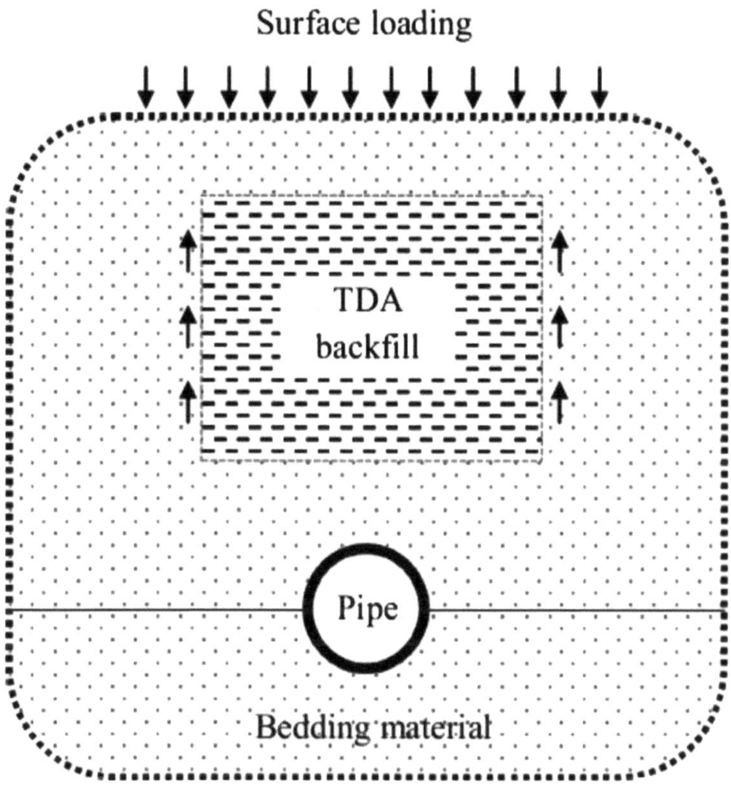

Fig. 1. Buried pipe and overlaid overlain by a tire-derived aggregate backfill (Source: Meguid and Youssef 2018)

Panjamani and Manohar (2012) investigated the shear strength characteristics, energy absorption capacity and brittleness index of sand-tire crumb mixtures (STCM). Initially, STCM samples were compressed and dilate upon shearing. With an increase in the rubber content for all sizes of tire crumb, this behavior increases. The friction angle will be controlled by the rubber grains in the sand-rubber mix. Cohesion intercept increases with increase in rubber content due to significant interlocking effects.

Ductility and energy absorption capacity of this mixture will be high. As rubber content increases, energy absorption capacity increases due to the increase in peak and ultimate stresses.

Reddy and Krishna (2015) stated that wall displacements were reduced remarkably by the addition of tire chips up to 30% by weight to sand (Fig. 2). There was a reduction because when tire chips content increases in the place of sand the unit weight will decrease. Due to the fact that tire chips have less specific gravity as compared to sand and a lower void ratio. It was observed that addition of tire chips in sand increases the shear strength properties like the angle of internal friction values. 50 to 60% earth pressures were reduced by the addition of 30% tire chips to sand. The response described above was due to the properties of tire chips like the lighter weight and high shear strength properties. The reduction of earth pressures and horizontal displacements implies a lower design requirement which implies lesser dimensions of the retaining wall.

Fig. 2. Sand-tire chip mixture (Source: Reddy et al. 2015)

Hataf and Rahimi (2006) carried out laboratory model tests to investigate the use of shredded waste tires as reinforcement to increase the bearing capacity of the soil. They said that the modeling of states of stress and strain in the reinforced sand during deformation and failure will be complex. The tensile force in the reinforcement can be divided into normal and tangential components to the shear plane. The normal component increases the confining stress on the failure plane and mobilizes the additional shear resistance in the sand. But the tangential component of tensile force directly resists the shear. The tensile force becomes greater than pull out resistance force of the reinforcements when tire shred length is smaller. For better performance, a minimum length of shred must be provided. If the length of tire shred becomes greater than the optimum length, then bearing capacity of reinforced soil decreases. This will be due to the fact that increasing the length will decrease area ratio. In general addition of tire shreds to sand increases the bearing capacity and limit reductions in post-peak resistance.

Tafreshi et al. (2012) carried studies on buried pipes placed in rubber-soil backfill. They estimated the combination of geocell reinforcement and rubber-soil mixture as a reinforcement. They concluded that rubber inclusion makes sand more compressible and causes improvement in bearing capacity and settlement of sand.

Marto et al. (2013) analyzed shear properties of sand-tire chip mixture using direct shear test apparatus. From the results, they observed that the shear stress increases with the number of tire chips in the mixture. But, after plotting graph between tire chips content and maximum shear stress for the different applied normal stresses, they noticed that the maximum shear stress decreases after 20% of tire chip content. From these findings, the researchers concluded that the normal stress, sand matrix unit weight, and tire content will affect the shear strength of the sand-tire chip mixture.

Lee et al. (2010) conducted a study to analyze the properties of rubber-sand mixtures and to understand the small strain behaviors based on their size ratios. After conducting Consolidation test using odometer, the researchers noticed that the shear wave velocity increases as the applied vertical stress and sand fraction in the mixture increases. They specified that the deformation of rubber particles will be 30 to 40 times more than pure sand particles. Irregularity and angularity of rubber particles increases as the rubber particle size decreases. As the sand fraction and the size difference between rubber-sand particles increases, the small strain stiffness increases rapidly.

Youwai and Bergado (2003) conducted triaxial compression tests on rubber tire-sand mixtures to study the strength and deformation characteristics. As the shear strength of the sand will be higher than the rubber-tire, the strength of the mixture will be more if the sand content is more than rubber-tire content. This mixture will be having lower compressibility when compared to the pure sand.

Ehsani et al. (2017) experimentally studied the behavior of soil-waste tire mixtures. Reduction of strength can be seen in the sand-tire waste mixture if tire content is more. Initial modulus of the sand-tire mixture and peak deviatoric stress reduces as tire content decreases in the mixture. By adding tire waste in the sand, the flexibility of sand increases. So, the sand-tire waste mixture can be used in flexible pavements and machine foundations.

Yadav and Tiwari (2017) carried out simple tests as per Indian Standard codes on clay-crumb rubber mixtures. On increasing the percentage of tire crumb mixture in clay, the maximum dry density and optimum moisture content decreased. The unconfined compressive and split tensile strengths decrease, when the tire crumb content crosses 5%.

Mashiri et al. (2013) studied the dynamic properties of sand-tire chip mixture by conducting cyclic triaxial tests. Rectangular shaped tire chips with a uniform width of 5 mm, aspect ratio 4 and varying depth from 3 to 7 mm, was mixed with beach sand in different proportions. The specimens of diameter 100 mm and height 200 mm were prepared for the consolidated undrained cyclic triaxial tests with optimum sand-tire chip mixture. From the test results, they observed that the shear modulus increases with the increase in the confining pressure and decreases with the increase in the number of cycles. But, as the number of cycles increases, the damping ratio started reducing its value. However, after first few cycles, the rate of decrease became insignificant. The damping ratio showed many variations with respect to shear strain. For the first few cycles, the damping ratio increases at all smaller shear strains, but after 40–60 cycles the damping ratio remains in between 9% to 10%. Whereas for larger shear strains, the damping ratio decreases after 10 to 20 cycles and remains in between 9% to 10%.

Noorzad and Raveshi (2017) compared the mechanical properties of unreinforced sand with sand reinforced with granulated and powder tire crumb, by performing triaxial tests. They varied the tire crumb (granulated & powdered) content from 0% to 30% in the mixture and conducted the tests. As the frictional coefficient reduces between the sand and tire crumb, the peak shear strength decreases. They observed that the increase of tire content in a mixture made the mixture to lose its peak shear strength and the granulated tire crumb – sand mixture provided high peak shear strength than powdered tire crumb – sand mixture. The ratio of tire crumb size to the mean diameter of sand plays a vital role in the volume change of reinforced sand due to shear deformation was greatly reduced by the addition of tire crumb to sand.

1.2 Waste Foundry Sand (WFS) as Backfill Material

Lee et al. (2001) carried research on the use of mixtures of WFS and fly ash as a controlled low strength material (CLSM). The main factors which contributed to use CLSM were flowability, strength, permeability, and settlement. To not compact the backfill material flowable fill can be used instead of conventional soil backfill. In this paper, they used green waste foundry sand, furane waste foundry sand and coated waste foundry sand. It was observed that when the specimens were cured for more than 15 days, they were sheared as negative pore pressure was developed. This was because the specimen showed an obvious shear plane, and separated parts of the specimen pushed out the membrane, leading to negative pore pressures. In general, the applied lateral earth pressure against the wall decreased up to 80% just 6 h after completion of the backfill, this was because of the good drainage at the top and bottom. The rate of the lateral earth pressure decrement highly depends on the type of waste foundry sand. Due to the rapid settlement of large particles and an increase in soil density at the bottom, the lateral earth pressure at the bottom of the retaining wall increased slightly. The lateral earth pressure at the bottom decreases or remains unchanged, despite the higher backfill, due to the cohesion of the mixtures with curing time. For the Green WFS mixture, due to segregation of the particles, it was observed that the lateral earth pressure at the bottom was larger.

1.3 Studies on Retaining Walls Under Static Loading

Tatlisoz et al. (1998) evaluated the mechanical properties of tire chips and soil-tire chip mixtures relevant to geosynthetic-reinforced earthworks in which Nicolon HS1150, a polyester woven geotextile was used as a geosynthetic. As the tire chip content increases, the apparent cohesion increases due to interlocking effect between sand and tire chips. The pullout effort required for sand-tire chip (STC) mixture is very low when compared to sand as the STC mixture is a light weight material. As the mixture is light in weight reinforced walls can be constructed with steep slopes. The friction between the sand and tire chips in STC mixture transmits the force from backfill to geosynthetic layer.

Lal and Mandal (2014) studied the behavior of cellular-reinforced fly-ash walls under strip loading using fly ash as an alternative backfill material in cellular-reinforced walls and utilizing locally available used wastewater bottles to prepare cellular reinforcements. In this study waste water bottles of two different diameters, 50 mm and 70 mm were used to reinforce the walls. The tests results showed that the fly ash can be used as an effective backfill material.

Reddy et al. (2015) observed that the accelerations were amplified more on top of the wall and the largest response acceleration of the model wall occurs in the case of sand alone. Tire chips will absorb the dynamic energy when dynamic force was applied to the model wall. After performing shake table tests on the model wall having a mixture of sand and tire chips as backfill, they observed that the backfill material can be effectively used for seismic resilience and provides better performance in adverse conditions like earthquakes.

Xie and Yang (2009) analyzed the interaction among wall, reinforced layers, and fill using the finite element software, MARC, and evaluated the stress distribution of the geocell flexible retaining wall. The horizontal displacement of reinforced layers is affected remarkably by its position. At small external loads, the reinforced layer at the top of the wall moves toward the fill behind the wall and the reinforced layers at other positions move away from the fill.

Chen and Chiu (2008) performed model tests on nine model geocell retaining walls (Fig. 3) to examine the effect of the geocells as a major material in retaining structures and the failure mechanism of these walls under surcharge. The lateral displacement of the wall increases with the increase of surcharge. The steep facing wall induces more displacement than the gentle facing wall, with the maximum value occurring at the top. Due to the steep wall face, the center of gravity of the wall becomes closer to the face of the wall. By this, the wall provides less resistance to overturning. When the wall tilted by the surcharge, the center of gravity shifts outward resulting in less resistance and then wall starts to fail by overturning. Due to this reason, the failure mode of the gravity-type wall includes horizontal sliding along the interface of geocell layers with an overturning of the wall. The maximum lateral displacement occurs at the top of the wall and the maximum settlement will be closer to the wall's face. The potential failure surface of an internally stabilized structure occurs within the reinforced zone, whereas an externally stabilized structure will have a failure within the soil mass, without penetrating through the structure. As geocell wall will be lighter, it produces more deformation and settlement than the gravity type.

1.4 Studies on Geocell Retaining Walls Under Dynamic Loading

Manju and Latha (2016) performed shake table tests on geocell retaining wall (Fig. 4) infilled with sand and gravel. They studied the effect of base acceleration, geocell material, infill material and basal reinforcement. The confinement effect and tensile strength of geocell play the main role in low strain seismic loading cases. The horizontal displacement was reduced in gravel infilled geocells than sand infilled. The overall stiffness of the wall was increased due to the higher stiffness provided by gravel infilled geocell. Basal reinforcement improves the confinement effect within the geocell facia as well as backfill soil and hence the overall performance improved with the

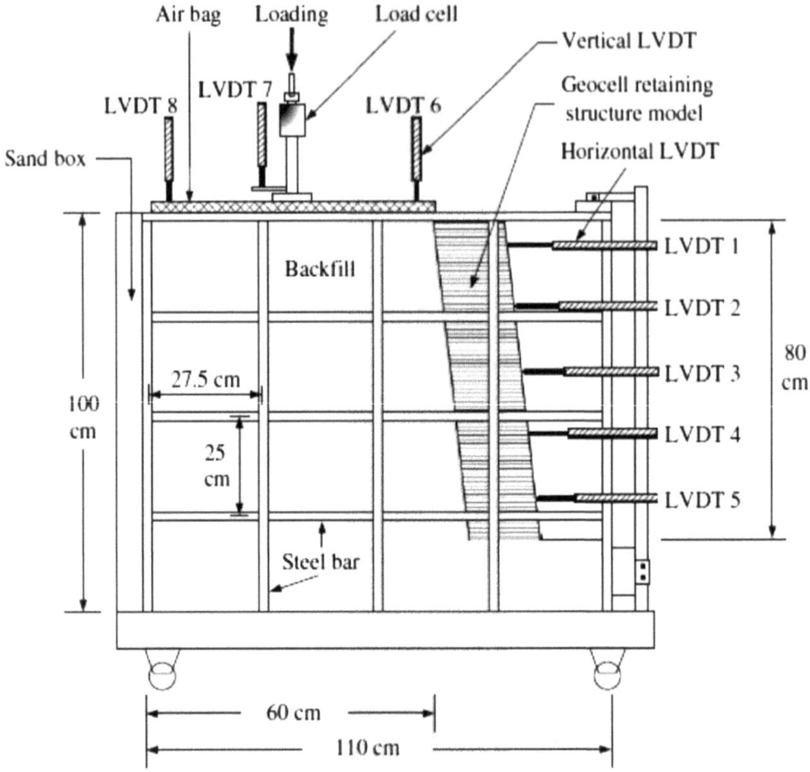

Fig. 3. Setup of instrumentation for geocell retaining wall-static loading test (Source: Chen and Chiu 2008)

inclusion of basal reinforcement. The test results showed that PVC geocells performed well in high seismic accelerations without any damage.

Latha and Manju (2018) carried out shake table tests to know the effect of the slope of the geocell wall on its seismic performance. The geocells were prepared with High-Density Polyethylene (HDPE) geonet and infilled with gravel. Two models, one vertical and another battered model were tested by varying base accelerations and frequencies. After performing shake table tests it was observed that the battered walls performed well than vertical walls. The improvement of stability of base of the battered wall was due to the shift of moment of inertia of pressure distribution at the back of the wall. Due to this, the stiffness of battered wall increased.

Ling et al. (2005) studied the behavior of reinforced soil under earthquake loading using finite element procedure. They observed that the lateral earth pressure behind the blocks was not affected by the different types of soil. The facing lateral displacement was affected by the behavior of soil, layout, and spacing of reinforcement. Ling et al. (2009) done experimental studies on the seismic response of five large-scale retaining walls having a geocell facing and investigated failure mechanism. Two of five walls were gravity retaining walls with geocell facing infilled with gravel and sand

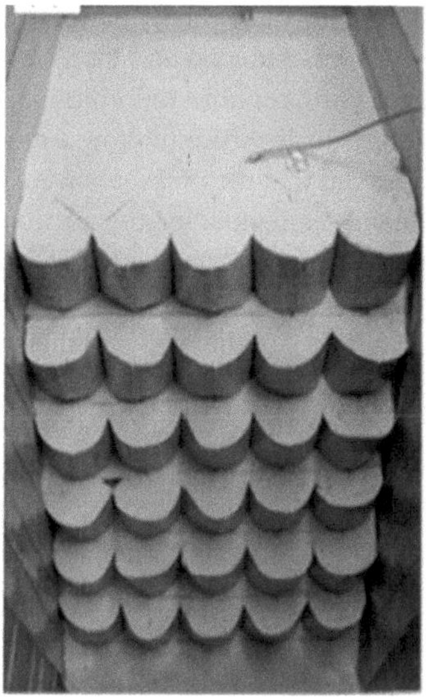

Fig. 4. PVC geocell model retaining wall (Source: Manju and Latha 2016)

respectively. The third geocell retaining wall was having geogrid as reinforcement. Other two walls were gravity retaining walls with geocell facing and with geocell reinforcement. It was said that, after first shaking, no surface crack or distress will be observed. A slight differential settlement between the facing and backfill with the cracks developed behind the backfill can be seen after second shaking. It was observed that the infill materials did not affect greatly the wall lateral displacements. The geocell confines the soil and prevents it from lateral deformation. Finally, they said that gravity walls made of geocells will be flexible than concrete gravity walls.

Leshchinsky et al. (2009) states that the small relative movement between geocell layers allows energy dissipation via friction while maintaining limited translational movement between layers. This is because of high friction between adjacent layers and strong interaction of the infill material within the HDPE cells. As the weight of the infilled geocells with gravel will be somewhat larger than the weight when infilled with sand, the interface frictional shear resistance along the geocell–sand increases. The geocell tensile resistance will contribute to the stability of the flexible structure.

Reddy and Krishna (2015) conducted studies on a retaining wall model with a conventional sandy backfill and another model in which tire chips were placed vertically as compressible inclusions. This was controlled by servo-hydraulic shaking table facility with a single degree of freedom and instrumented with pressure sensors, displacement transducers, and accelerometers at different locations. They observed that the accelerations were amplified at top elevations and the displacements increased significantly with the increasing base acceleration of input motion. During the dynamic excitation, incremental lateral earth pressures were observed along the height of the wall. These lateral earth pressures exhibited an increasing trend with an increase in base acceleration. With an increase in compressible inclusion thickness, the displacements significantly decreased. Due to the lightweight and compressible nature of the tire chips, the soil thrust on the wall was reduced and thereby the wall displacements were reduced. With the inclusion of compressible tire chips just behind the wall, the sand backfill will tend to move more in the backfill for a model wall with thicker compressible inclusions. They observed that the pressures were higher at the bottom and the earth pressures were significantly affected by the tire chips inclusions. Due to the compressible nature of the tire chips at the wall–soil interface, higher acceleration amplifications can be observed. The compressible inclusions bring the soil to the active state when these were placed in-between the wall and the backfill soil. Reduction of the lateral earth pressure and displacement makes a lower design requirement possible, which entails smaller dimensions of the retaining wall with reduced material cost. The additional advantage of providing tire chips as a compressible inclusion near the wall facing is the provision of drainage, which will reduce the pore pressure effects on the wall.

1.5 Conclusion and Future Scope

- From past studies, it can be concluded that ductility and energy absorption capacity of the mixture of sand and tire crumbs will be the most promising one.
- The flexible geocells are characterized by a ductile stress-strain response. They serve as dampers by absorbing and dissipating the dynamic energy.
- The gravity retaining walls and reinforced concrete slabs, are rigid and brittle. When these rigid structures are subjected to vibrations which are generated by earthquakes, they sustain and transfer the vibrations with little attenuation or even amplify the load.
- Reinforcing geocell layers, provide more stability against dynamic loading, in addition to the stabilization of the backfill. So, the infilled geocell with the mixture of sand and tire waste/waste foundry sand and tire waste can become a good seismic resistant material.

References

Chen, R.H., Chiu, Y.M.: Model tests of geocell retaining structures. Geotext. Geomembr. **26**, 56–70 (2008)

Edinçliler, A., Ayhan, V.: Influence of tire fiber inclusions on the shear strength of sand. Geosynth. Int. **17**(4), 183–192 (2010)

Ehsani, M., Shariatmadari, N., Mirhosseini, S.M.: Soil Waste Tier-Sharif Iran.Pdf, vol. 24, pp. 65–71 (2017)

Feng, Z.Y., Sutter, K.G.: Dynamic properties of granulated rubber sand mixtures. Geotech. Test. J. **23**(3), 338–344 (2000)

Hataf, N., Rahimi, M.M.: Experimental investigation of bearing capacity of sand reinforced with randomly distributed tire shreds. Constr. Build. Mater. **20**, 910–916 (2006)

Lal, R.R., Mandal, J.N.: The behavior of cellular-reinforced fly-ash walls under strip loading. J. Hazard., Toxic, Radioact. Waste, ASCE **18**, 45–55 (2014)

Latha, G.M., Manju, G.S.: Effect of facing slope on the seismic response of geocell walls. Geotech. Eng. **49**(1), 73–83 (2018)

Lee, C., Truong, Q.H., Lee, W., Lee, J.-S.: Characteristics of rubber-sand particle mixtures according to size ratio. J. Mater. Civ. Eng. **22**(4), 323–331 (2010)

Lee, K., Cho, J., Salgado, R., Lee, I.: Retaining wall model test with waste foundry sand mixture backfill. Geotech. Test. J. **24**, 401–408 (2001)

Leshchinsky, D., Ling, H.I., Wang, J.P., Rosen, A., Mohri, Y.: Equivalent seismic coefficient in geocell retention systems. Geotext. Geomembr. **27**, 9–18 (2009)

Ling, H.I., Leshchinsky, D., Wang, J.P., Mohri, Y., Rosen, A.: Seismic response of geocell retaining walls: experimental studies. J. Geotech. Geoenvironmental Eng. ASCE **135**, 519–524 (2009)

Ling, H.I., Liu, H., Mohri, Y.: Parametric studies of the behavior of reinforced soil retaining walls under earthquake loading. J. Eng. Mechanics. ASCE **131**(10), 1056–1065 (2005)

Mashiri, M.S., Sheikh, M.N., Vinod, J.S., Tsang, H.: Dynamic properties of sand-tyre chips mixtures. In: Australian Earthquake Engineering Society Conference, 15–17 November 2013, Hobart, pp. 1–8 (2013)

Meguid, M.A., Youssef, T.A.: Experimental investigation of the earth pressure distribution on buried pipes backfilled with tire-derived aggregate. Transp. Geotech. **14**, 117–125 (2018)

Manju, G.S., Latha, G.M.: Innovative cellular confinement systems for retaining walls. In: Geo-Chicago 2016. ASCE (2016)

Marto, A., Latifi, N., Moradi, R., Oghabi, M., Zolfeghari, S.Y.: Shear properties of sand – tire chips mixtures. Electron. J. Geotech. Eng. **18**, 325–334 (2013)

Ni, P., Qin, X., Yi, Y.: Numerical study of earth pressures on rigid pipes with tire-derived aggregate inclusions. Geosynth. Int. 1–45 (2018)

Noorzad, R., Raveshi, M.: Mechanical behavior of waste tire crumbs-sand mixtures determined by triaxial tests. Geotech. Geol. Eng. **35**(4), 1793–1802 (2017)

Panjamani, A., Manohar, D.R.: Shear strength characteristics and static response of sand-tire crumb mixtures for seismic isolation. In: International Conference on Sustainable Built Environment, Kandy, Sri Lanka (2012)

Reddy, S.B., Krishna, A.M.: Recycled tire chips mixed with sand as lightweight backfill material in retaining wall applications: an experimental investigation. Int. J. Geosynth. Ground Eng. **3**, 1–11 (2015)

Reddy, S.B., Krishna, A.M., Dasakai, M.S.: Seismic resilience of retaining walls backfilled with sand–tire chips mixtures. In: International Workshop on Geotechnics for Resilient Infrastructure, The Second Japan-India Workshop in Geotechnical Engineering (2015)

Tafreshi, S.M., Mehrjardi, G.T., Dawson, A.R.: Buried pipes in rubbersoil backfilled trenches under cyclic loading. J. Geotech. Geoenvironmental Eng., ASCE **138**(11), 1346–1356 (2012)

Tatlisoz, N., Edil, T.B., Benson, C.H.: Interaction between reinforcing geosynthetics and soil-tire chip mixtures. J. Geotech. Geoenvironmental Eng. ASCE **24**(11), 1109–1119 (1998)

Xie, Y., Yang, X.: Characteristics of a new-type geocell flexible retaining wall. J. Mater. Civ. Eng. ASCE **21**(4), 171–175 (2009)

Yadav, J.S., Tiwari, S.K.: Influence of crumb rubber on the geotechnical properties of clayey soil. Environ., Dev. Sustain. 1–22 (2017)

Youwai, S., Bergado, D.T.: Strength and deformation characteristics of shredded rubber tire-sand mixtures. Can. Geotech. J. **40**(2), 254–264 (2003)

Analysis and Solution of Containment Stabilization in Gabions Held at the BR-060 Margins in the Municipality of Guapó/Goiás, Brazil

Rideci Farias[1(✉)], Haroldo Paranhos[1(✉)], Itamar de Souza Bezerra[2(✉)], Silvio Lourenço da Silva Filho[2(✉)], Ranieri Araújo Farias Dias[3(✉)], and Roberto Pimentel de Sousa Júnior[4(✉)]

[1] Reforsolo Engineering, Brasília, Brazil
rideci.reforsolo@gmail.com, reforsolo@gmail.com
[2] Maccaferri, Goiânia, Brazil
Itamar@maccaferri.com,
silvio.lourenco@maccaferri.com.br
[3] Federal Univesity of Pará, Belém, Brazil
ranierileislie@yahoo.com.br
[4] University of Brasília, UnB, Brasília, Brazil
eng.robertopimentel@gmail.com

Abstract. This work presents the rupture in a road cut slope with an approximate extension of 100 m, located on the banks of the BR-060 in the urban stretch of the municipality of Guapó/GO. The studies consisted in the execution of a program of field incursions to verify the geotechnical conditions in the indicated section, collection of soil samples, execution of laboratory tests, analysis of slope stabilities and the consequent proposition of stabilization using the structure of containment in gabions of 6.0 m height, 2: 1 slope (H: V) and subsequent planting of grass with protection against surface erosion. The solution adopted in gabions to contain the damaged section proved to be advantageous, considering the economic aspect of other solutions, and its executive simplicity provided speed to the executive process, allowing execution in the rainy season without any kind of intercurrence. In order to verify the current conditions of the implanted system, a visit was made at the beginning of 2017 to the place where the intervention took place and it was found that the implanted works are consistent with the expected operation.

1 Introduction

With the implementation and (or) adaptation of highways, many times, there is a need to it is necessary to cut in certain sections that cause considerable slopes that deserve special attention of the geotechnical part. In this way, this work presents the study of a 100-meter-long stretch of road with cut slopes with approximate heights of 10 (ten) meters located on the banks of BR-060, metropolitan region of Goiânia, in the

H. Ameen and T. Sorour (Eds.): GeoMEast 2018, SUCI, pp. 182–194, 2019.
https://doi.org/10.1007/978-3-030-01929-7_13

Fig. 1. Location of the municipality of Guapó/GO in relation to the capital Goiânia/GO.

municipality of Guapó/GO. Figure 1 shows the location of the municipality of Guapó/GO in relation to the capital Goiânia, and Fig. 2 shows the BR-060 stretch in the municipality of Guapó/GO.

This section is an integral part of the duplication of BR-060, a work of the federal government under the management of the transport ministry and executed by the National Department of Transport Infrastructure (DNIT), which extends from Goiânia to the municipality of Jataí/GO.

Figure 3 shows the situation of the slope studied before the deflagration of the rupture.

Fig. 2. Featured area of BR-060 in Guapó/GO.

Fig. 3. View of the duplication of BR-060 in the Guapó/GO section.

2 Breakthrough in the Cut - BR-060

Figures 4, 5, 6 and 7 present a photographic survey carried out on October 26, 2011 with the conditions of the section that had rupture, located in the area of Stake 593. It can be seen in the photos that the lateral slope to the Highway was in the process of degradation and/or progressive instabilization that required urgent interventions in order to mitigate the current problem. It was also observed that the problem was

Fig. 4. View of the stretch that has ruptured.

Fig. 5. View of the stretch that has ruptured.

Fig. 6. View of the section that has ruptured.

Fig. 7. View of the stretch that has ruptured.

aggravated by the percolation of water from concrete pipes, causing excess pore pressure, reducing the shear strength of the soil.

Table 1, below, shows the section verified with the altitudes of the crest and the foot of the slopes, as well as the heights of the same ones. In terms of geometry, in the verified section, the slopes present heights up to approximately 10 m, as evidenced in the sections analyzed and shown in this work.

Table 1. Approximate extensions of each cut studied.

Talude			
Estaca	Altitude (m)		Altura (m)
	Crista	PÉ	
590	719,728	711,320	8,408
591	720,672	712,091	8,581
592	721,430	712,861	8,569
593	721,901	713,632	8,269
594	722,914	714,402	8,512
595	723,609	714,624	8,985
596	724,357	714,818	9,539
597	725,159	715,498	9,661
598	725,953	716,101	9,852
599	726,749	717,735	9,014
600	727,639	719,275	8,364
601	728,318	720,359	7,959

Figure 8 shows the profile of the broken slope between stakes 593 and 595. The upstream and downstream stretches did not break, however, they presented erosions at different evolutionary stages, indicating a high susceptibility to new damage, the decision was taken to study an area for containment that contemplated the stretch subject to possible instability.

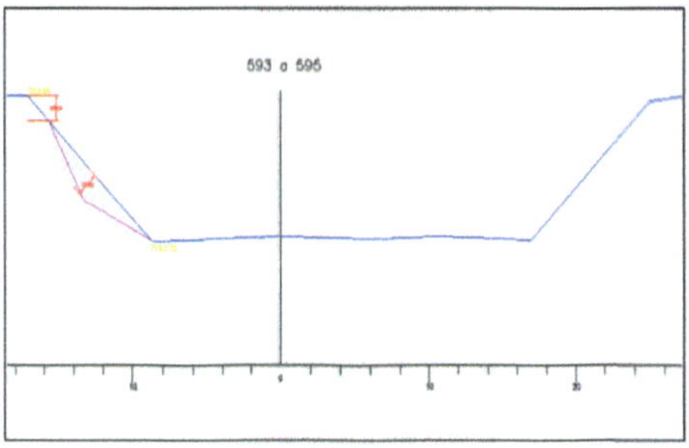

Fig. 8. Cross section of cut after rupture.

3 Geotechnical Characterization of the Area

In order to determine the parameters necessary to check the stability of the slopes, three undisturbed blocks were collected from the ruptured section (Stake 593 to 595). The samples were named Sample 1, 2 and 3. In tactile terms, Sample 1 corresponds to a red to variegated soil. This soil is predominant in the ruptured section. Sample 2 corresponds to a variegated sandy clay and Sample 3 corresponds to a sparse variegated sandstone.

The three Soil Samples were used to perform the tests of natural moisture, granulometry (sieving and sedimentation), real grain density, liquidity limits and plasticity. However, for Sample 1, as it is the most representative in the ruptured section, as well as other stretches, the direct shear tests were carried out in order to define the resistance parameters of this sample.

For the direct shear tests, the specimens were tested in the natural condition in order to verify the stability of the slopes in this condition, but also in the flooded condition aiming to verify the possible critical mechanical behavior of the soil in the rainy season and (or) a possible saturation caused by other factors.

According to INMET (2012), the cumulative rainfall in the period was 190 mm, corresponding to the beginning of the rainy season in the region.

Table 2 shows the identification of Sample 1 as well as the summary of the results of the geotechnical tests carried out and parameters determined.

Table 2. Summary of the results of the geotechnical tests carried out and parameters determined for Sample

Amostra 1		
Peso específico natural	γ_{nat} (kN/m^3)	16,71
Umidade natural	w (%)	12,9
Coesão efetiva (solo natural)	C (kPa)	16,61
Ângulo de atrito efetivo (solo natural)	ϕ' (graus)	28,0
Coesão efetiva (solo inundado)	C (kPa)	5,1

(*continued*)

Table 2. (*continued*)

Amostra 1		
Ângulo de atrito efetivo (solo inundado)	ϕ'(graus)	27,0
Peso específico aparente seco (kN/m^3)	γ_d (kN/m^3)	14,8
Peso específico dos sólidos (ou dos Grãos)	γ_s (g/cm^3)	2,780
Limite de liquidez	(%)	45,1
Limite de plasticidade	(%)	36,2
Índice de plasticidade	(%)	8,9
Fração argila	(%)	35,6
Fração silte	(%)	41,1
Fração areia	(%)	23,2

4 Analysis of Effective Stabilities

Stability analyzes were performed using the SLOPE/W and MACSTARS®2000 programs, considering the soil parameters in the natural and flooded condition, according to the direct shear tests, with a slope crest overload of 20 kPa in both situations and finally the stability of the solution considering the use of the gabions was calculated. Figures 9, 10, 11 and 12 show some of the analyzes performed and Table 3 shows a summary of the results of these analyzes with the safety factors obtained.

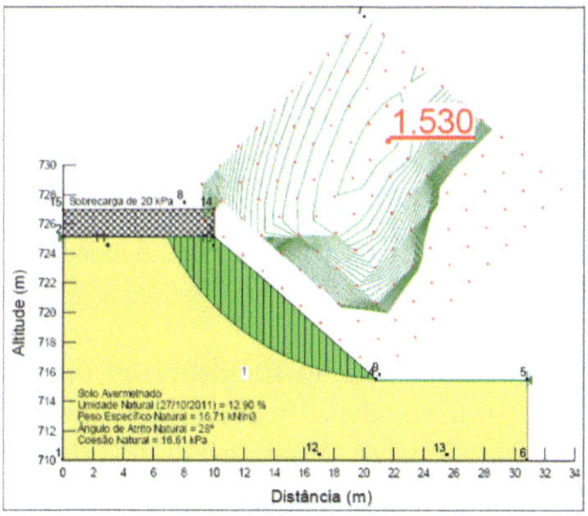

Fig. 9. Stability slope stability of stake 597 (parameters of soil in natural condition with overload of 20 kPa.). Bishop method.

The results obtained between the methods of Bishop and Morgenstern-Price, summarized in Table 3, were very close and consistent with the actual situation of the work studied. It is noticed that, being the soil in the saturated condition, the slope is potentially unstable and thus with the leakage of the pipe and the intensification of the rainy period in the region, there was the outbreak of the instabilization process.

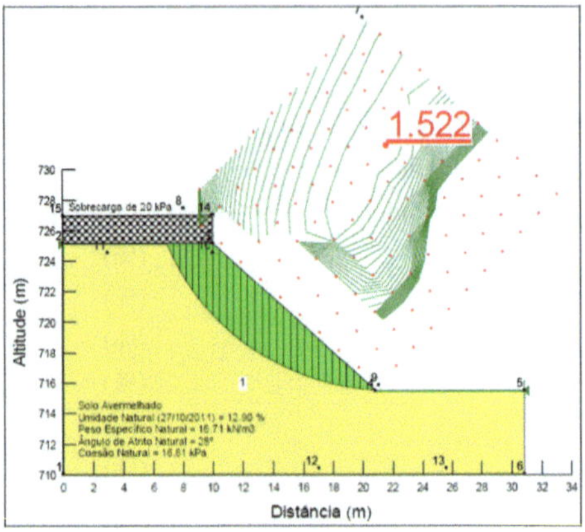

Fig. 10. Slope stability of stake 597 (parameters of soil in natural condition with overload of 20 kPa). Morgenstern-Price.

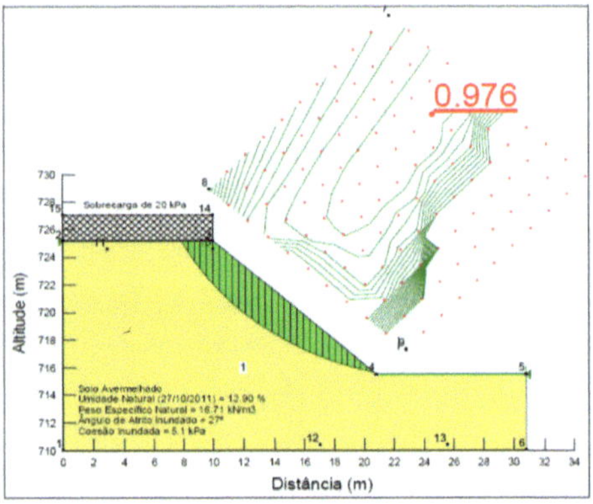

Fig. 11. Slope stability of stake 597 (soil parameters in flooded condition with overload of 20 kPa). Bishop method.

The analyzes were carried out without simulation of the water table because this situation was not detected for the analyzed section. Only the parameters for the soil in the flooded condition were considered. However, it should be noted that in the rainy season, there is the natural flow of rainwater by the existing slopes so that there may be some infiltration into the soil. It should also be noted that in some situations the soil of the massif is in the unsaturated condition, in the zone above the water level, which can be considered in the analysis by the inclusion of the resistance parameter Φb. However, this situation was not considered by the fact of the current rainy season. As a form of exemplification, the equation of shear strength is presented as:

$$\tau = c' + (\sigma - u_a)\tan\phi' + (u_a - u_w)\tan\phi^b \quad (1)$$

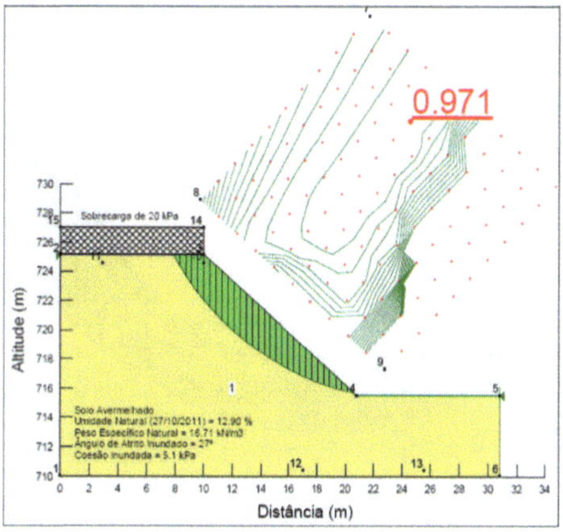

Fig. 12. Stability slope stability of stake 597 (soil parameters in flooded condition with overload of 20 kPa). Morgenstern-price.

Table 3. Summary of the results of the analyzes performed with the safety factors obtained.

Condições	Método-fator de segurança (FS)		Média (FS)
	Bishop	Morgenstern-price	
Com parâmetros do solo natural - Com sobrecarga de 20 kPa na crista do talude	1,53	1,522	1,526
Com parâmetros do solo inundado - Com sobrecarga de 20 kPa na crista do talude	0,976	0,971	0,9735

Being:

(σ − ua) = tensão normal líquida
ua = pressão de ar;
(ua − uw) = sucção mátrica.

5 Adjusted Stabilization Solution

In order to meet the needs of the geotechnical, economic and executive aspects of the work in question, a containment structure was defined as a gabion retention structure with a height of 6.0 m and a tipping (2H: 1 V), followed by planting grasses aiming at protection against surface erosion.

Figures 13 and 14 show the stability analyzes studied for the case in question already contemplating the insertion of the solution adopted in gabions, where in addition to the evaluation of the overall stability of the structure, the safety coefficients were also evaluated for sliding, tipping and pressures in the foundation.

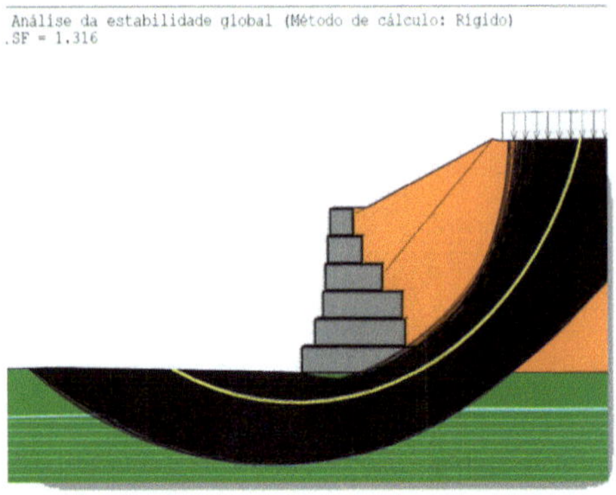

Fig. 13. Global stability analysis, contemplating solution in gabions, with FS: 1.31 (parameters of soil in the flooded condition with an overload of 20 kPa).

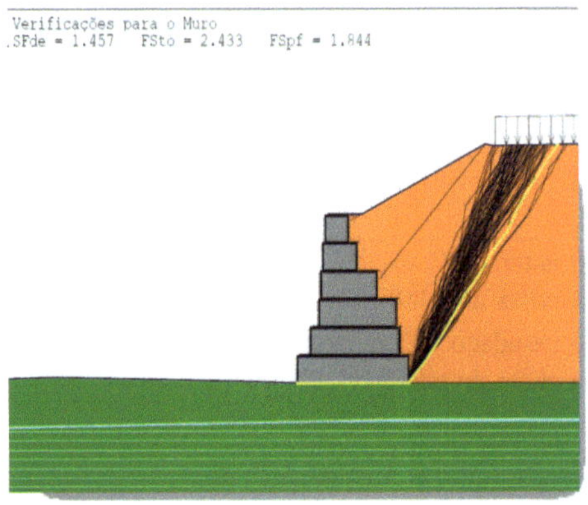

Fig. 14. Slip stability/tipping and foundation analysis, considering solution in gabions, with FSde = 1.45; FSto = 2.43; FSpf = 1.84 (soil parameters in flooded condition with overload of 20 kPa).

The stability analyzes presented in Figs. 13 and 14 show that the gabion solution is efficient for the present case and could be adopted as a form of stabilization.

Stability was achieved with six layers of gabions, the first with 5.0 m length, the second with 4.0 m, the third with 3.5 m, the fourth with 2.5 m, the fifth with 1.5 m and the last layer with 1.0 m in length. At the end of the structure, a nonwoven geotextile, MacTex N40.2, with a tensile strength of 10 kN/m was used, with the main function of promoting the filtration process, facilitating drainage and preventing the leakage of fine material into the interior of the boxes.

The gabions are modular elements, with varying shapes and are made from double-twisted hexagonal mesh fabrics, typically being filled with stones of suitable granulometry and sewn together. Thus, in line with good engineering practice are intended for the solution of geotechnical, hydraulic and erosion control problems. The solution has as main differentials, the high resistance and flexibility of the metallic meshes, great drainage potential due to the storage of the stones inside the boxes, versatility, low environmental impact with fast integration to the surrounding environment and last but not least, are considerably economical when compared to similar solutions.

6 Execution of the Solution

After cleaning the ground with the removal of the ruptured material, the assembly and mooring of the gabions were started, obeying the design section. The filler material should have good specific gravity and must not be friable, guaranteeing the behavior and the expected strength for the structure. The filling of the cartons should be in stages, as shown in Fig. 15.

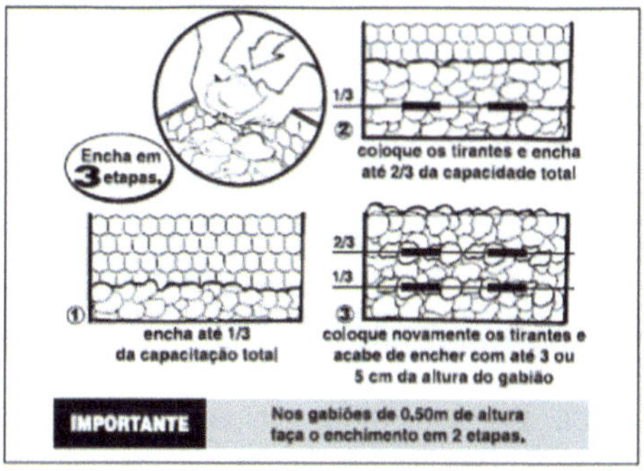

Fig. 15. Steps for filling the box gabions and placing the tie rods (barros 2005).

Figures 16, 17 and 18 show the execution of the gabion system, and in Fig. 19 the finished wall is shown. Subsequently, in Figs. 20 and 21 the stage of the work is presented at the beginning of 2017.

Fig. 16. Execution of gabion restraint system.

Fig. 17. Execution of gabion restraint system.

Fig. 18. Execution of the gabion restraint system.

Fig. 19. Finished wall able to receive the top-up shredding.

Fig. 20. Steps for filling the box gabions and placing the tie rods (Google Earth 2017).

Fig. 21. Steps for filling the box gabions and placing the tie rods (Google Earth 2017).

7 Conclusions

This study sought to present and analyze the use of techniques in gabion to recover a stretch that suffered a slope rupture. The studies carried out showed that:

(a) The lateral slope to the Highway was in the process of degradation and/or progressive instabilization that required urgent interventions in order to mitigate the current problem. It was also observed that the problem was aggravated by the percolation of water from concrete pipes, causing excess pore pressure, reducing the shear strength of the soil;
(b) The solution adopted in gabions proved to be advantageous, considering the economic aspect in front of other solutions and its executive simplicity provided speed to the executive process, allowing the execution in the rainy season without any kind of intercurrence;
(c) In order to verify the current conditions of the implanted system, the site where the intervention occurred in the beginning of 2017 was visited and it was verified that the implanted works are consistent with the expected operation.

Acknowledgment. The National Department of Transport Infrastructure (DNIT), Maccaferri do Brasil Ltda., Reforsolo Engenharia Ltda., Universidade Católica de Brasília (UCB), University Center of Brasília (UniCEUB) and the Institute of Higher Education Planalto with important contributions that have made possible the accomplishment of this work.

References

INMET. Instituto Nacional de Meteorologia (2012). http://www.inmet.gov.br. Accessed 02 Feb 2012

Barros, P.L.A.: Obras De Contenção - Manual Técnico, Maccaferri do Brasil Ltda, São Paulo (2005)

Google Earth. http://www.google.com/earth/index.html. Accessed 13 July 2017

Solutions of Recovery in Slopes and Access to the Transmission Line in 500 KV in the Section Located in the Municipalities of Colinas do Sul and Niquelândia, State of Goiás, Brazil

Rideci Farias[1(✉)], Haroldo Paranhos[1], Elson Oliveira de Almeida[2], Maicon Vitor Oliveira[3], Erani Bastos[4], Marcelo Oliveira[4], José Augusto de A. Lopes[4], and Udo Gebrath[5]

[1] Reforsolo Engineering, Brasília, Brazil
rideci.reforsolo@gmail.com, reforsolo@gmail.com
[2] UniCEUB, Brasilia, Brazil
elsul0@hotmail.com
[3] MMJ Engineering, Brasília, Brazil
mmjengenharia@gmail.com
[4] Dossel Ambiental Cosunsulting, Brasília, Brazil
{erani,marcelo,augusto}@dosselambiental.com.br
[5] State Grid Brazil Holding S.A., Rio de Janeiro, Brazil

Abstract. The implementation and operation of a linear enterprise as an energy transmission line involves a series of activities that, depending on the nature of the terrain, can cause diverse impacts to the environment. Among environmental concerns in construction are slope stability, erosion control, sediment generation from excavations and land movements, but also the need to recover areas affected by the work along the route. Thus, in places susceptible to erosive processes, where access openings, areas for the installation of towers and maneuvers, cuts and embankments are necessary, it is necessary to adopt preventive and (or) corrective measures in order to avoid the onset of new erosions and (or) correct existing ones. Another concern is the stability of cutting slopes and embankments that can give rise to landslide susceptible to landslides, both by exposure of unprotected material and by the stability of the soil mass itself. Therefore, in many situations there is a need for practices of containment structures, compaction of landfills, surface protection of soil, execution of surface and/or subsurface drainage systems, among other measures, to minimize possible impacts. These measures also aim to preserve existing facilities, as well as the enterprise itself and the environment. The main justification is the need for recovery and stabilization of accesses, slopes and terraced areas, avoiding possible damages to the soils, vegetation, hydrographic system, sources and the integrity of the enterprise. Therefore, this article presents the procedures adopted to try to return to the areas that have been interfered with, the environmental characteristics that were before and (or) adapt them to the new reality, and also necessary for the operation and maintenance of a transmission line.

H. Ameen and T. Sorour (Eds.): GeoMEast 2018, SUCI, pp. 195–210, 2019.
https://doi.org/10.1007/978-3-030-01929-7_14

1 Introduction

Generally the phase of implementation of a linear enterprise as a transmission line (LT) of electric energy is the one that causes the greatest impacts to the environment. For this, actions are needed that produce practical and quick effects to the recovery of the areas affected by the installation of the undertaking and (or) due to the action of nature, but also by anthropic issues. In this context, it adopts criteria and devices to be considered in order to recover, protect and stabilize the affected sites and seek balance with the surrounding areas aiming at the continuity of LT operation and maintenance. In order to do so, the aim is to present a diagnosis of the impacts of these projects, as well as the usual procedures adopted to try to return to the areas that were interfered with, the environmental characteristics they were before and (or) adapt them to the new reality, and maintenance of a LT installed in the States of Goiás, Federal District and Minas Gerais.

2 Characteristics and Location of the Enterprise

The LT in question is of 500 kV, with extension of 675 km, and begins in the municipality of Colinas do Sul/GO and follows to the municipality of Araguari/MG. The route is monitored in substations in the municipalities of Luziânia/GO and Paracatu/MG. From Luziânia, heading for the Northwest, towards Samambaia/DF, operated by Furnas and located in Recanto das Emas Administrative Region in the Federal District. In this article, emphasizing the northern part of the LT, in the Municipalities of Colinas do Sul and Niquelândia, State of Goiás.

3 Physical Characteristics of Areas of Influence

In general terms, the enterprise is inserted in the Central Plateau, involving the area of plateaus, which constitutes the watershed of the Tocantins, Paraná and São Francisco river basins. Composed of areas dissected by the constituents of these basins and flattened areas of the Alto Tocantins Depression. These reliefs are supported by rocks of the crystalline basement of Archean and Proterozoic age, sedimentary and Mesozoic igneous rocks and Cenozoic detritus-lateral coverings.

The region is inserted in the Morphoclimatic Domain of the Tropical Chapadões covered by Cerrados and penetrated by Forest Galleries that presents high temperatures and well defined periods of rain. The summer months: December, January and February are the rainiest, with the driest periods being the months of May to September.

It is verified that the main restrictions regarding the physical environment, in the Area of Direct Influence (AID), are the mountains and cliffs with greater intensity in the northern section. These formations intercept in some passages transversally to the AID. Figure 1 shows the mapping of the location of the enterprise, Figs. 2 and 3 show an overview of the typical relief in the northern section of the transmission line, and

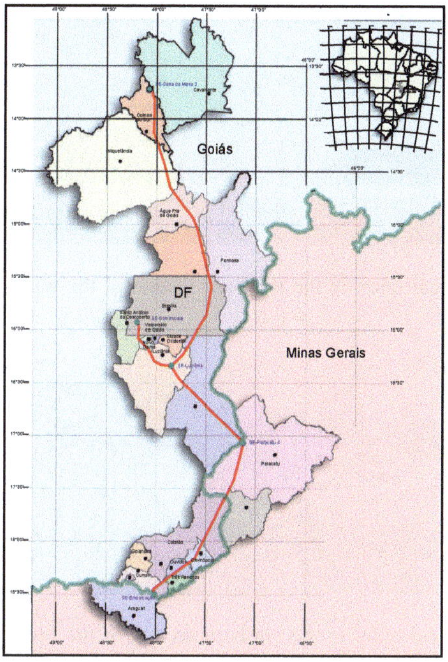

Fig. 1. Location map of the development.

Figs. 4 and 5 show a characteristic slope, as well as occurring natural landslides, even without anthropogenic interventions.

Another constraint linked to the physical environment is the regions with little availability of access. These regions often coincide with reliefs of escarpments and escarpments, but in other cases they occur on reliefs less rugged, but with high drainage density, where the need to construct the access roads generates significant environmental impacts. It is also emphasized that among the conditions of the physical environment are the dissected reliefs that develop from the edges of the board. In general this type of formation presents progressively greater declivity as it approaches plains, which makes the almost longitudinal trajectory, relatively unfavorable. In general, the AID drainage network carved rather dissected reliefs, whose gaps are reduced from north to south. Such relief results in chapadões interspersed by more or less abrupt slopes, observed in the lower third of the northern stretch and in the other stretches. In the upper two-thirds of the northern section, there is the formation that gives name to the section, formed by some chapadões, the "mesas", interspersed by high altitude saws. The escarpments of these saws present relatively flat intrusions, the pediments, where the installation of the LT encountered minor difficulties. The following are the types of lands observed and crossed by the project with the main geotechnical characteristics. (a) Fluvial Plains/Veredas: Presence of high water table and risks of annual floods. Soft soils with precarious stability of the excavation walls and the possibility of settlements; (b) Summit ramps: they may have laminar erosion

Fig. 2. Typical relief in the northern section of the transmission line.

Fig. 3. Typical relief in the northern section of the transmission line.

Fig. 4. Natural sliding view.

and furrows. High intensity erosion at the edges of the plate; (c) Hill and Ramps: Possibility of laminar erosion and furrows in the areas of earthworks and drainage of service roads and excavation of foundations. Localized sedimentation of river channels. Presence of bumps in the ground and outcrops of rock that may hinder excavations. (d) Morrotes, hills and hills and steep hills: Presence of bumps in the ground and outcrops of rock that makes excavations difficult. Risk of localized landslides due to exposure of soil/rock contact in saturated or upwelling areas and descaling on slope or slope surface. High soil erosion.

Fig. 5. Natural sliding view.

4 Diagnosis of the Situation of Access to Towers of the Transmission Line

The main environmental liabilities found in the field survey are related to drainage, erosive processes, material deposition and slope instability. In this work, the slopes are emphasized. For the development of this work, a broad photographic record was carried out in the months of August 2016 and January 2017 in order to subsidize field and office work. This record was made by means of photographic cameras and capture of images obtained from the drones with the purpose of indicating technically feasible and effective solutions for the recovery of the areas, emphasizing that in function of the local dynamics, such solutions can be adapted or even modified in order to seek the balance with the environment of insertion of the enterprise. Thus, even with the actions already defined, it is quite possible that they may be adapted, since when they are deployed in the field, they are accompanied by specialized professionals in order to seek the highest degree of efficiency in the interventions. Figures 6, 7, 8, 9, 10, 11, 12, 13, 14 and 15 show the opening of access routes to the enterprise that can cause instability in slopes.

Fig. 6. Opening of access road to LT tower for installation and maintenance.

Fig. 7. Opening of access road to LT tower for installation and maintenance.

Fig. 8. Opening of access road to LT tower for installation and maintenance.

Fig. 9. Opening of access road to LT tower for installation and maintenance.

Fig. 10. Opening of access road to LT tower for installation and maintenance.

Fig. 11. Opening of access road to LT tower for installation and maintenance.

Fig. 12. Opening of access road to LT tower for installation and maintenance.

Fig. 13. Opening of access road to the LT tower for installation and maintenance.

5 Main Actions Adopted for Recovery of Degraded Areas

The actions of the operation phase, comply with the Brazilian technical norms and the pertinent legislation, as well as the practice of engineering. The activities related to the enterprise and the environment are monitored and monitored permanently, mainly in the northern part of the enterprise in which are verified areas susceptible to mass movement and erosive processes. Among the main actions taken in the correction and (or) recovery aiming, in particular, the stability of the slopes, are:

Fig. 14. Opening of access road to LT tower for installation and maintenance.

Fig. 15. Opening of access road to LT tower for installation and maintenance.

(a) Correction/Adequacy of the access road bed;
(b) Construction/Recovery of ridges;
(c) Reconforming/restoring slopes;
(d) Reconstruction/Reconstitution of manholes;
(e) Repair of wooden structures (palisades);
(f) Construction of a gabion with handstand;
(g) Slopes in slopes;
(h) Execution of gutters;
(i) Construction of hydraulic ladders;
(j) Installation of biomass and revegetation of slope;
(k) Implantation of metal screen in slopes;
(l) Execution of containment system with sacaria;
(m) Application of Cal-Jet in slopes;
(n) Use of sowing on slopes.

The following are the main techniques adopted for the recovery of degraded areas, emphasizing the stability of slopes with correction systems, drainage, protenção or even the recovery of slopes.

5.1 Correction/Adequacy of the Access Road Bed

In order for the runway to properly exercise its paper, slope adjustments are made to allow safe rideability, but also to protect the platform from deterioration. Correction activities consist of regularization of the access bed, leaving them slightly inclined ($\approx$4%) to the base of the cut, especially in the situations of curves when they coincide with canyons, increasing the safety of the road and facilitating the capture of water through the channel of the cut or gutter of the access road.

5.2 Camps Construction/Recovery

The ridges will be constructed following the technique of terraces, formed by the combination of a canal (ditch) and a ridge (earth mound or dyke), according to Fig. 16, at intervals scaled transversely to the access road bed, forming a barrier and conducting the runoff to the water outlet.

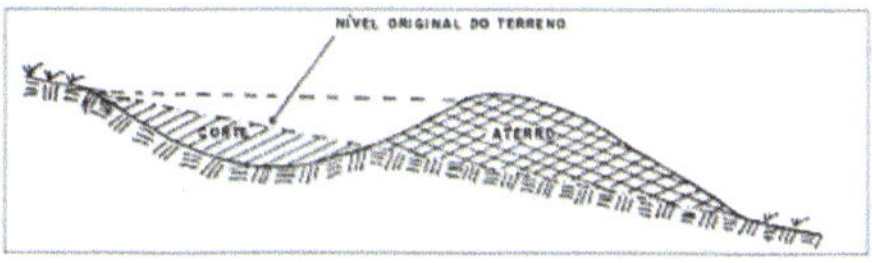

Fig. 16. Example of simple ridge cut.

5.3 Slope Reconforming/Remodeling

Process carried out by machinery and manually to correct and/or prevent erosions and landslides of loose materials that may fall on the runway, with danger to traffic or risk of obstruction of drainage systems.

5.4 Wood Structures (Palisades)

Palisades are bulkheads that aim at the retention of sediments from surface runoff. In the case in question, they are constructed with wooden props and boards (Figs. 17 and 18).

Fig. 17. Palisade used for sediment retention.

Fig. 18. Palisade used for sediment retention.

5.5 Construction of Gabion with Local Stone

Armed structures (Figs. 19 and 20), flexible, draining, durable and resistant. In the case in question local stones are used in order to minimize costs.

Fig. 19. Slope containment system with gabion with local stone.

Fig. 20. Slope containment system with gabion with local handstand.

5.6 Biomass Installation and Slope Revegetation

The biomass has basically as raw material the coconut fiber and agricultural straw, used in several practices of erosion control. The main characteristics are the permeability, surface protection of the soil against surface runoff and other intemperes, environmental integration because they are degradable and environment conducive to the

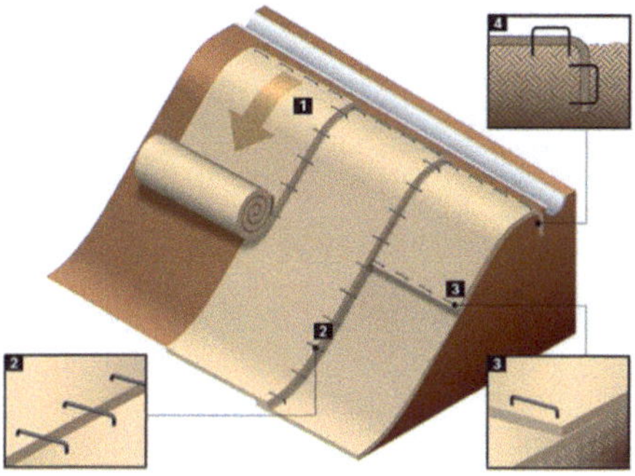

Fig. 21. Detail of the Biomantas installation.

Fig. 22. Consorciation of grasses and legumes, associated with biomanta.

development of plant species such as grasses and legumes. Figures 21 and 23 show basically the technique used in the undertaking in question (Fig. 22).

5.7 Metal Screen Implantation

The high resistance mesh is a technique used to complement the slope containment for slides and falls of rock blocks, or as a biomass fixer (Figs. 23, 24, 25 and 26).

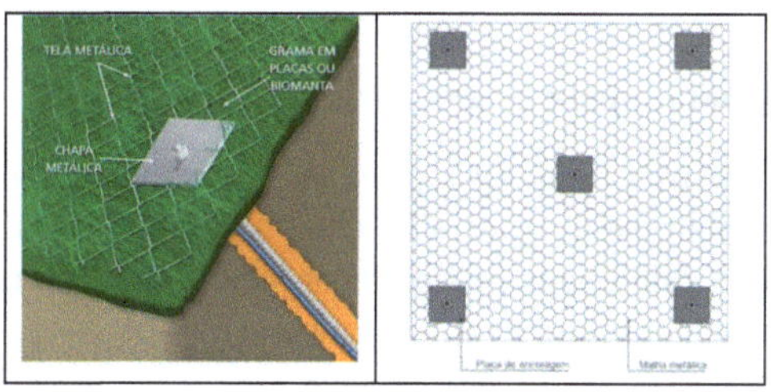

Fig. 23. Example of resistance plate fixation with plate (Rather used for rock material).

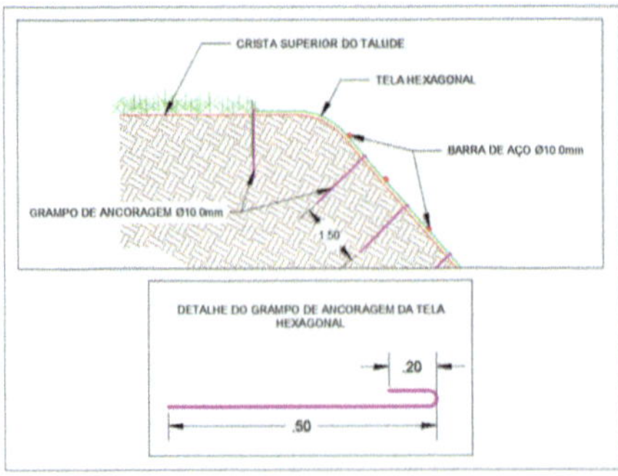

Fig. 24. Example of rebar fixation (Enough used in soil).

Fig. 25. Application of metal screen to avoid material fall.

Fig. 26. Application of biomanta and metallic screen for fixing.

5.8 Execution of Containment System with Sacaria

The riparian system, referred to as rip-rap, is an alternative technique for slope stabilization and slope containment, generally used to obturate erosions (Figs. 27, 28 and 29). Generally with raffia bags or line bags, when the soil-cement mixture solidifies, the bags are no longer required in structural terms of the containment work.

Fig. 27. Example of use of drainage for erosion fill.

Fig. 28. Use of drainage to protect slopes.

Fig. 29. Use of drainage to protect slopes.

5.9 Application of Cal-Jet on Slopes

A technique developed in 2009 by Geologist Álvaro Rodrigues Santos, is subsidized in the spraying of lime slurry with fixative binders on the soil surfaces to be protected (Figs. 30 and 31). The great asset of the Cal-Jet technique is ensured by the combination of the following attributes: good performance, low cost, practicality of application, high efficiency in application (m2/day/operator) and great durability. The Cal-Jet technique uses hydrated lime for paint produced and marketed by several brands.

Adaptations in the hydrated lime produced for mortar (plaster) may in the future allow the use of this lime in the Cal-Jet technique, which would be interesting given its lower cost. Lime paint (lime + water) is not a solution, it is a suspension because the

Fig. 30. Application of Cal-Jet with metal screen to protect slopes.

Fig. 31. Cal-Jet with metal screen for slope protection.

lime is practically insoluble in water. In this way, lime paint must be permanently shaken during application to avoid sedimentation.

Note: The design and development of the Cal-Jet technique is the sole responsibility of its creator, being the holder of its patent (INPI-PI0603110-2). However, your brush use by everyone who comes to make use of it is totally released from royalty payments or any other types of property rights reimbursements. Only the commitment of strict compliance with the recommendations expressed in its Execution Manual is requested.

5.10 Sowing

According to the technical communiqué of Embrapa, n° 105 (2007), the styler-large field does not need previous preparation of the soil, as long as the amount of seeds applied is increased. The ideal sowing should have depth of 1 to 3 cm, because superficially or at greater depths there is no good establishment of the plant.

The area to be re-vegetated has a high slope and, following the recommendation of the Technical Manual for the Recovery of Degraded Areas of IBAMA (1990) for these conditions is used the technique of hydroseminate and in addition, manual (or haul). For both, there is a need for shallow seed coverage after sowing.

5.10.1 Water Hydration

It consists of sowing herbaceous species in aquo-pasteous environment, composed of substances that aid in the fixation and growth of plants and in the retention of moisture.

It requires adequate equipment, consisting of a motor-pump and tank with stirring blades, as well as a hose or cannon for dispersion of the mixture (IBAMA 1990).

5.10.2 Manual Seeding

Manual sowing (Figs. 32 and 33) is performed in addition to the hydroseminate, when deemed necessary. After opening the pits/grooves the seeds are homogeneously poured along the area, considering the indicated proportion of kilograms of seeds per square meter.

Fig. 32. Manual seeding.

Fig. 33. Manual seeding.

6 Conclusions

This article tried to present the procedures adopted to try to return to the areas that have been interfered with, the environmental characteristics that were before and (or) adapt them to the new reality, and also necessary to the operation and maintenance of a transmission line, and currently, year of 2017, the executive processes are underway and due monitoring is done so that future publications can show the behavior of each solution used.

Acknowledgment. State Grid Brasil Transmission, Dossel Ambiental Consultoria e Projetos Ltda., Reforsolo Engenharia Ltda., MMJ Engenharia Ltda., Universidade Católica de Brasília (UCB), University Center of Brasília (UniCEUB) and Instituto de Ensino Superior Planalto with contributions that made this work possible.

References

EMBRAPA, Comunicado Técnico 105. ISSN 1516-9308. Campo Grande, MS (2007)
IBAMA, Manual Técnico de Recuperação de Áreas Degradadas. Brasília, DF (1990)
Consultas a diversos "sites" da internet realizadas no ano de 2017 com vistas a familiarização aos processos executivos das variadas técnicas indicadas e utilizadas em campo

Evaluation of the Containment Curtain Staking Influence on Underground Water Flow Through the Seep/W Software Utilization

Guilherme Ribeiro Paiva[1(✉)], Rideci Farias[2], Haroldo Paranhos[2], Ranieri Araújo Farias Dias[3], Roberto Pimentel de Sousa Júnior[4], and Lucélia Lisboa Ribeiro[5]

[1] Civil Engineering, IESplan, Brasília, Brazil
guilherme933@gmail.com
[2] Reforsolo Engineering Ltd., Brasília, Brazil
rideci.reforsolo@gmail.com, reforsolo@gmail.com
[3] UFPA, Belém, Brazil
ranierileislie@yahoo.com.br
[4] UnB, Brasília, Brazil
eng.robertopimentel@gmail.com
[5] Civil Engineering, UniCEUB, Brasília, Brazil
lulu.lb@hotmail.com

Abstract. With the increasing urbanization and land use, engineers are forced to seek solutions in underground works. In order to enable the implementation of this methodology, it is often used a solo containment system, being it through retaining walls, diaphragm walls, or as usual as it is in Brazil, concrete pile walls. When there is the necessity to achieve greater depths, a common problem occurs while reaching the water level: the outcrop of the water table. Therefore, it is necessary to use, in addition to a containment system, a groundwater lowering system; being it temporary or permanent - or both. The combination of these two systems, containment and groundwater drawdown, then becomes necessary for the execution of underground works in these exceptional cases. The study of the underground water flow behavior and their permanent or temporary drawdown, as well as the influence promoted by concrete pile walls, are topics with low academic production in Brazilian bibliographies. This article aims to deepen the knowledge about the greater efficiency and economy provided by the use of these two systems together by analyzing and producing information in the construction field.

1 Introduction

The objective of this paper is to analyze the influence of the containment curtain staking on underground water flow (juxtaposed stakes) using the Seep/W calculation software of the Canadian company Geo-Slope International Ltd.

It will be shown a real case work in a building of 17000 m^2, located in Olímpia, Sao Paulo, Brazil.

H. Ameen and T. Sorour (Eds.): GeoMEast 2018, SUCI, pp. 211–224, 2019.
https://doi.org/10.1007/978-3-030-01929-7_15

In this case, the water table was, at certain points, 0.60 m deep (Figs. 1, 2, 3, 4, 5 and 6), being necessary to dig an underground pavement to a depth of 3.50 m. Additional block foundations, underground reservoirs and lifts shafts were also built; totaling 5.50 m depth.

Fig. 1. Situation before lowering system

Therefore, a temporary water table lowering system was designed through drainage wells.

After the complete execution of the system and the excavation in accordance with the project quota, it was noticed that the few wells located near the containment curtain of the project were not as necessary for the effective lowering of the water table as other wells in the project.

The purpose of this project is to understand, with the help of a software called Seep/W, how the containment curtain line influenced the flow of underground water by naturally lowering it only through its execution.

By understanding how the containment curtain affected the water table, it will be possible to develop a more succinct project with the same efficiency but with reduced costs, thus improving the cost-benefit rate – the main goal of an engineering project.

Fig. 2. Situation before lowering system.

Fig. 3. Comparison situation before lowering system

Fig. 4. Comparison situation after lowering system

2 Development

2.1 Methodology

All the basic and theoretical concepts used in this study were extracted from research sources, such as books and technical journals. For additional information in the general development of this research, specialized websites were also consulted.

The lowering of the water table is an issue rarely studied in Brazil, therefore, few sources of national bibliographic information were found. The work that was taken into account the most was the calculation system and downgrade proposed by Alonso, U. R. (2007) *Rebaixamento Temporário de Aquíferos*, São Paulo, Oficina de Textos.

3 Materials

The methodology of this article consists in analyzing, through the use of the Seep/W software, the influence of the containment curtain stacking on the underground water flow.

This consideration became known after the execution service of a groundwater lowering system project in a hotel company in Olímpia, São Paulo, Brazil.

Fig. 5. Situation after lowering system

This project was made by means of drainage wells specifically calculated with the local soil data obtained.

After the project was fully implemented by the business owner, a peculiarity was noticed: some drainage wells, located near the containment structure, were not useful.

Then, the main question that gave purpose to the writing of this article emerged: how and how much do the containment curtain structures impact on underground water flow?

All materials and procedures used in the investigation of such problems will be listed in the following topics.

3.1 Building Data

The project to be studied is a hotel and water park complex, in the city of Olímpia, in the state of São Paulo, Brazil.

The hotel and water park complex has 04 towers, each with 17 floors, an underground pavement completely buried and a ground floor half buried.

3.2 Local Soil Data

The soil survey report was made by means of a Standard Penetration Test (SPT) in November 2012.

Fig. 6. Situation after lowering system

A total of 16 SPT holes were made in an area of 17000 m^2 in order to obtain a report which characterized the soil and observed the water level in the corresponding month of its execution.

Based on the values found in the SPT report, two 3D modeling pictures with the coordinates XY of each SPT hole were made, one for the water lever (Fig. 7) and other for the impenetrable level (Fig. 8).

Analyzing the SPT report, it was observed that the water level had a peak of outcropping in the upper right corner of the building, influenced also by the geological characteristics of the place. In this point, the impenetrable level is very close to the surface, about 7 m deep. In some points near this critical region there were water levels very close to 0.60 m from the terrain surface.

In general, the NA had an average value of 3.40 m deep.

The soils characterized in the SPT report were: sandy clay in the first 3 m; clay silt from the third to the sixth meter; silt sand from the sixth until the fourteenth meter, in this case, the end of the SPT.

3.3 Temporary Water Table Lowering System Data

Initially, the method proposed to lowering system calculation consisted in dividing the business area into 4 parts of 50 × 50 m.

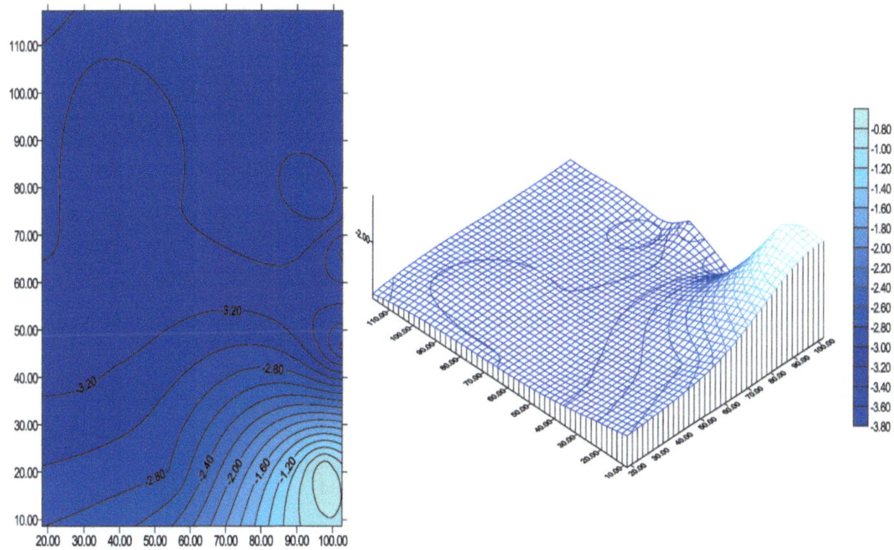

Fig. 7. Water level digital modeling

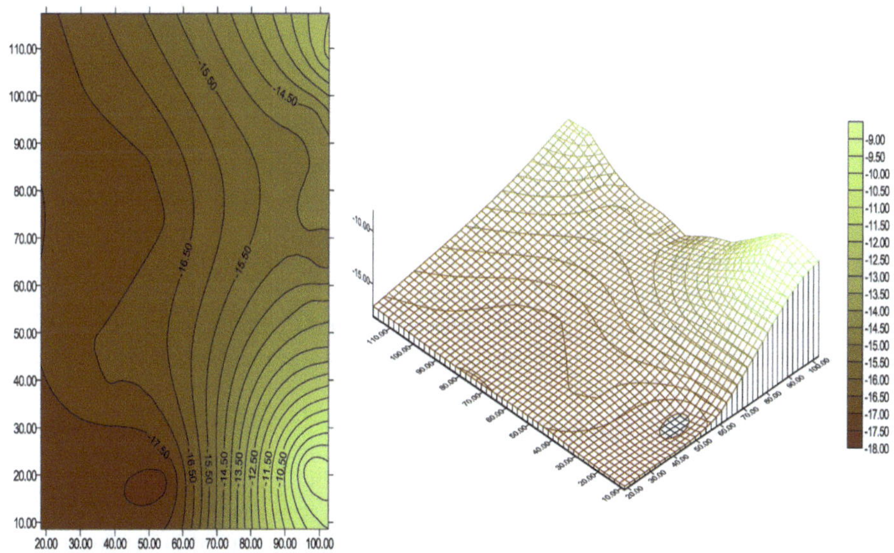

Fig. 8. Impenetrable level digital modeling

After calculations, as recommended by Alonso's formulas (2007), there was a need for each part to place 08 drainage wells with a diameter of 0.50 m and a depth of 12 m (Table 1). It was 4 parts, and then the total number of drainage wells was 32 units.

To assure the correct work of the system, an executive methodology was adopted. This methodology was carried out as described below.

Table 1. Original temporary water table lowering system for each quarter

Well diameter (meters)	0.50
Well radius (meters)	0.25
Soil permeability average (meters/second)	0.00001
Pit width (meters)	50
Pit length (meters)	50
Limited area (square meters)	2500
Well equivalent radius (meters)	28.20
Water level height as of the rock (meters)	12
Height of water table lowered to rock (meters)	7
Influence ray of drainage well (meters)	106.1
Total system flow (cubic meters/second)	0.011
Number of drainage wells (units)	8
Distance between drainage wells (meters)	25
Flow of each drainage well (square meters/second)	0.0015
Drainage well filtration length (meters)	2.10
Flow rate in the filter (meters/second)	0.00047
Groundwater reduction (meters)	3.60
Water height inside the drainage well to the rock (meters)	3.40
Drainage well length (meters)	10.70
Extra safe additions (meters)	1.50
Total drainage well length (meters)	12.20

I. Implementation of provisional wells, with a diameter of 1 m and depth of 2.50 m strategically positioned to help draining some water found in excess on the surface, enabling the ground to start next steps.
II. Implementation of trenches with width of 0.60 m and depth of 1 m in all of internal perimeter of the building land.
III. Implementation of drainage wells after steps I and II, which could take from 10 to 20 days depending on the conditions presented in the work environment.

3.4 Seep/W Software Application

With the software Seep/W a computational analysis about the influence of stacking of the containment curtain on the underground water flow was made.

A graphic based on CAD (Computer Assisted Design) was used in the analysis. It was drawn from the software Seep/W to create a digital model of the case.

The software worked with the building real dimensions. Data regarding the soil, hydrology and water levels variation found in the SPT report were also used.

In addition, the containment curtain stacking line location has also been carefully added to the software.

After organizing all data properly, it was added to the contour conditions of the software, trying to keep the digital model as close as possible to the real situation experienced.

3.4.1 Inserting Data and Parameters into the Seep/W Software

The digital model started in Seep/W software by inserting some project data as size, width and length of the building and so, an initial analysis plan was created (Fig. 9).

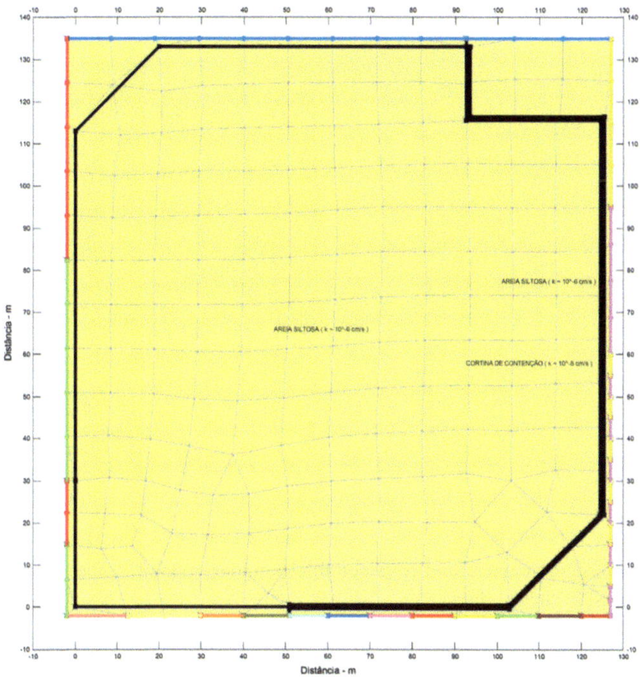

Fig. 9. Analysis plant set up and configured. The ticker line represents the containment curtain

On average, from the 6 m depth, only silt sand was found until the impenetrable surface. Since the analysis was made using a flat view perspective (a view from the top), it was not possible to define different layers of materials on top of each other, so only the silt sand was adopted, presenting permeability $k = 1 \times 10^{-6}$ cm/sec. As the software only accepts input in meters, it was set as $k = 1 \times 10^{-8}$ m/s.

After that, a fictitious region of 1 m wide was created to simulate the line of containment curtain. In the real situation, the containment curtain line is arranged by concrete piles of 0.40 m spaced every 0.70 m.

A weighted average was calculated and the permeability $k = 1 \times 10^{-8}$ cm/s (for software $k = 1 \times 10^{-10}$ m/s) in the region signaled by the containment curtain.

After the materials were defined, the limit parameters of the thickness of the soil layer were added, which means the boundary of the impenetrable level was defined. From the data obtained in the SPT report, 20 m were adopted as average thickness.

Then, the hydrological parameters were inserted into the software. The results of the groundwater level curves in digital model were in accordance with the results and graphs obtained by the SPT report (Fig. 10).

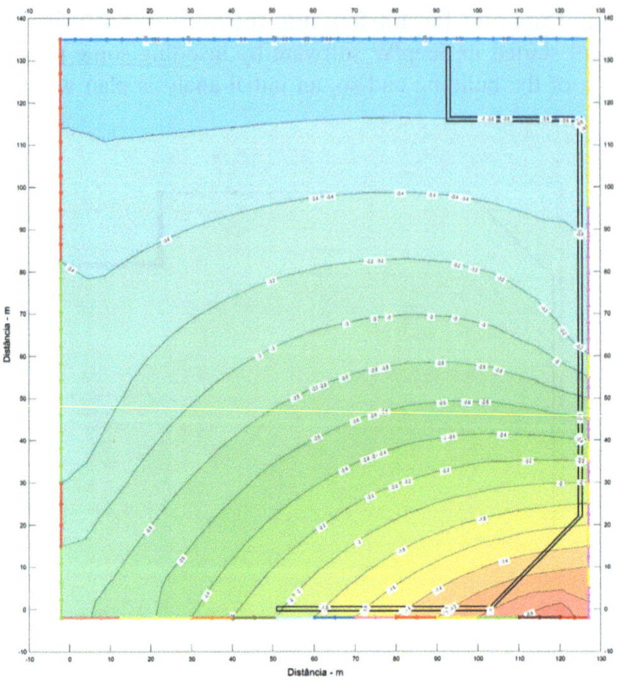

Fig. 10. Hydrologic parameters and impenetrable boundary added to the software so that they match the actual SPT survey data

3.4.2 Generating Analysis in Seep/W Software

After obtaining the necessary data, the software started running the calculation process.

The results obtained can be seen in Fig. 11.

From these software results, 4 sections in plan view were chosen (Fig. 12) and then charts were drawn up to analyze the water level variation as it moved away from the containment curtain line (Figs. 13 and 14).

4 Analysis of Results

It could be possible to infer from Fig. 14 that, after inserting the containment curtain element, there was a general lowering of the water level.

When observing the points of the lower face of the hotel, where the curtain makes a curve and seems to "embrace" the building, before the insertion of the staking representative element, the water was being maintained at 1.20 m and, after insertion of the representative element, the water table was lowered to 2.20 m, thus representing 1 m of effective lowering in this area.

However, as it progressed northward, becoming more distant from the containment curtain, this difference gradually decreased until it stabilized at the groundwater level

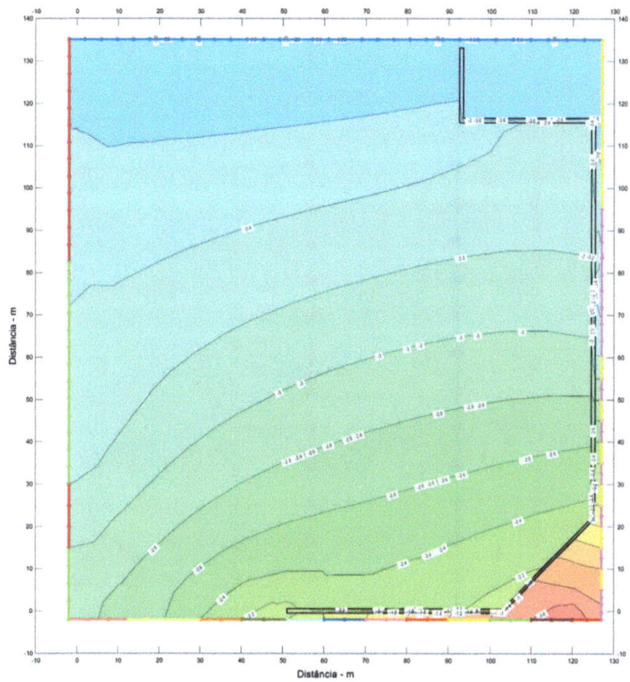

Fig. 11. Groundwater contours recessed after insertion of the retaining curtain element

before the insertion of the representative containment element, being it in about 80 m away from the curtain.

From the results obtained, it was possible to recalculate the water table lowering system previously designed.

5 System Recalculating and Analytical Comparisons

Based on the results obtained through the software, the height value of the water table was modified and the system was recalculated. No other modifications, rather than this new water table level analysis, was made, and so it was possible to compare the new system sizing based only in the correct theoretical containment curtain influence in the water level.

By the same means, the new dimensioning was executed, dividing the business into 4 parts of 50 × 50 m adopting wells with the same diameter of about 0.50 m.

The interaction of the formulas resulted in wells with a slightly lower depth than initially calculated.

The new total depth of the wells resulted in 11.50 m. The previous system was 12.20 m.

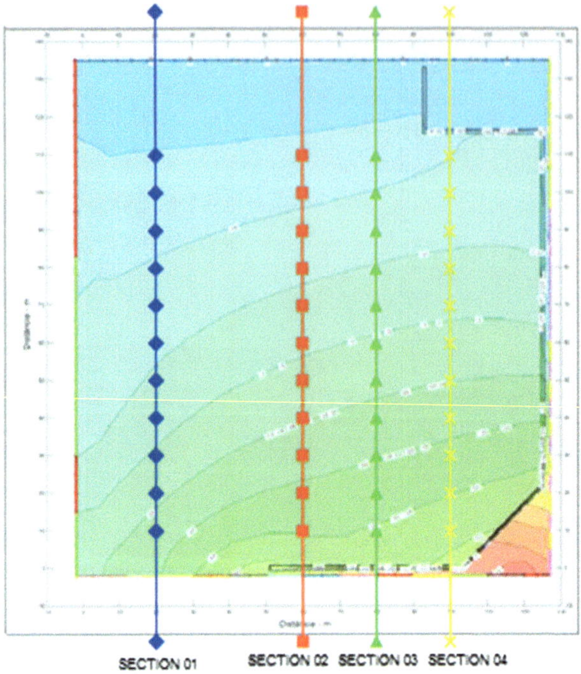

Fig. 12. Plan of analysis sections

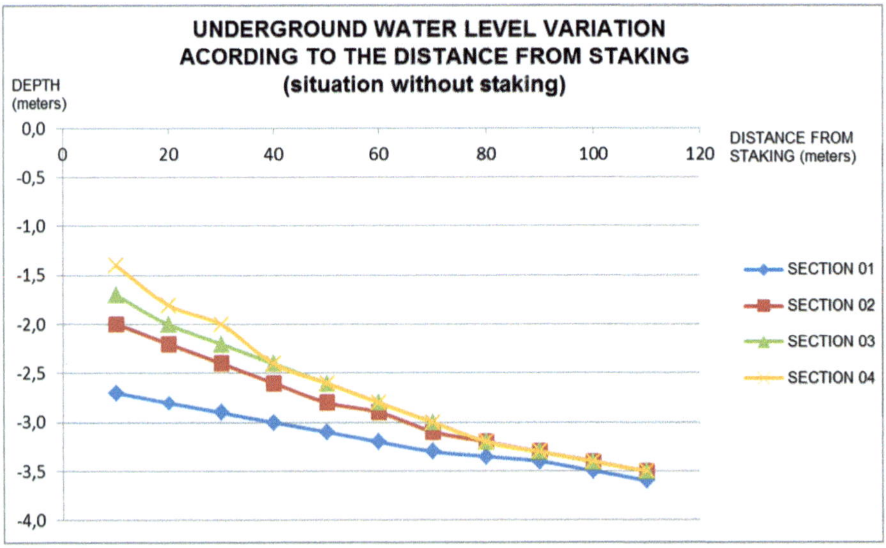

Fig. 13. Variation of underground water level without staking

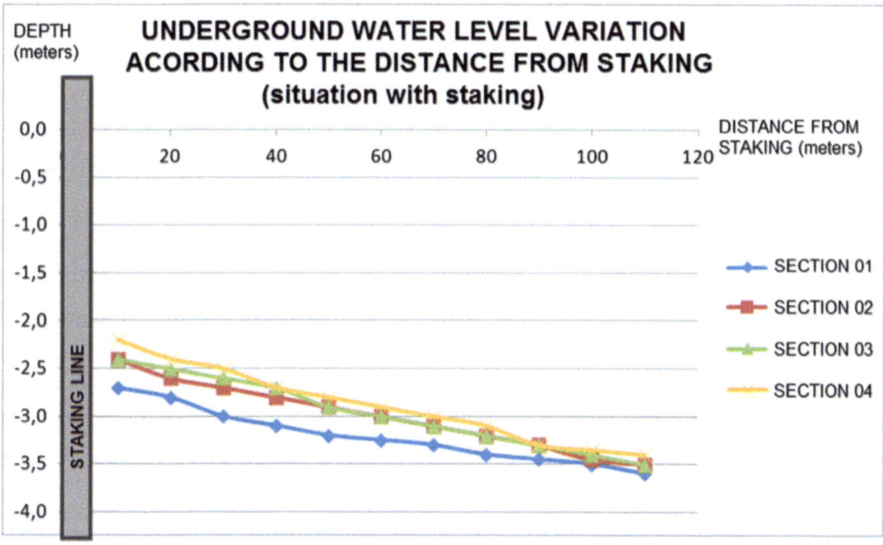

Fig. 14. Variation of underground water level with staking

The simple fact that the reduction in the height of the water level led to a diminution in the depth of the drainage wells at 0.70 m shows that this real interpretation about the containment curtain influence can generate huge savings in materials, time and workers to the business constructor.

The resizing results may be followed in Table 2.

Table 2. Recalculated water table lowering system for each quarter

Well diameter (meters)	0.50
Well radius (meters)	0.25
Soil permeability average (meters/second)	0.00001
Pit width (meters)	50
Pit length (meters)	50
Limited area (square meters)	2500
Well equivalent radius (meters)	28.20
Water level height as of the rock (meters)	11.50
Height of water table lowered to rock (meters)	7
Influence ray of drainage well (meters)	95.50
Total system flow (cubic meters/second)	0.011
Number of drainage wells (units)	8
Distance between drainage wells (meters)	25
Flow of each drainage well (square meters/second)	0.0015
Drainage well filtration length (meters)	2
Flow rate in the filter (meters/second)	0.00047
Groundwater reduction (meters)	3.50

(continued)

Table 2. (*continued*)

Water height inside the drainage well to the rock (meters)	3.50
Drainage well length (meters)	10
Extra safe additions (meters)	1.50
Total drainage well length (meters)	11.50

6 Conclusions

According to the facts presented in this work, it was concluded that the stacking of the containment curtain only contributed to the groundwater lowering, close to the staking process and not throughout the construction, as imagined at the beginning of this study.

However, it could be perceived that in regions where the staking "embraced" the building, creating a kind of region that we can call "protected" by the stacking line, the water table had a considerably large drawdown from 0.8 to 1.0 m approximately regarding its original level. But as the staking ceased to "embrace" the building, its influence on the effective lowering of the water table was very small, being it around 0.40 to 0.60 m.

The limit of containment curtain staking influence was seen up to 80 m from the stacking line, since there was no effect on the water flow.

It is necessary to point out that this limit distance is not an universal parameter, but it is possible for similar cases to be based on it. Other very important observation is that each case will require a new analysis in the Seep/W software – that means adding new parameters of the soil, water level, containment curtain density and, mainly, its geometry and layout settings.

It can also be concluded that the fact that some drainage wells were not useful could have occurred for other reasons rather than the containment curtain stacking line influence. For example: the bad execution of the well *in situ*; or perhaps because the well has been located in a region surrounded by a soil with lower permeability; the influence of the adjacent wells can also be considered and, finally; the water may simply have infiltrated towards regions of higher hydraulic gradient where the percolation was easier, thus not reaching to be captured by these tested wells.

Acknowledgments. I am grateful to the teachers and civil engineers D. Sc. Rideci Farias and M. Sc. Haroldo Paranhos for the theoretical help regarding the geotechnical issues used during the elaboration of the article until its publishing.

I am grateful to the civil engineers Ranieri Farias, Roberto Pimentel and Lucélia Lisboa for the help with publishing.

I am grateful to the English teacher Thiago Farias for the guidance in the translation of this article to its English version.

Reference

Alonso, U.R.: Rebaixamento Temporário de Aquíferos. São Paulo. Oficina de Textos (2007)

Author Index

H. Ameen and T. Sorour (Eds.): GeoMEast 2018, SUCI, p. 225, 2019.
https://doi.org/10.1007/978-3-030-01929-7

MIX
Papier aus verantwortungsvollen Quellen
Paper from responsible sources
FSC® C105338

If you have any concerns about our products,
you can contact us on
ProductSafety@springernature.com

In case Publisher is established outside the EU,
the EU authorized representative is:
Springer Nature Customer Service Center GmbH
Europaplatz 3, 69115 Heidelberg, Germany

Printed by Libri Plureos GmbH
in Hamburg, Germany